Springer Theses

Recognizing Outstanding Ph.D. Research

Aims and Scope

The series “Springer Theses” brings together a selection of the very best Ph.D. theses from around the world and across the physical sciences. Nominated and endorsed by two recognized specialists, each published volume has been selected for its scientific excellence and the high impact of its contents for the pertinent field of research. For greater accessibility to non-specialists, the published versions include an extended introduction, as well as a foreword by the student’s supervisor explaining the special relevance of the work for the field. As a whole, the series will provide a valuable resource both for newcomers to the research fields described, and for other scientists seeking detailed background information on special questions. Finally, it provides an accredited documentation of the valuable contributions made by today’s younger generation of scientists.

Theses may be nominated for publication in this series by heads of department at internationally leading universities or institutes and should fulfill all of the following criteria

- They must be written in good English.
- The topic should fall within the confines of Chemistry, Physics, Earth Sciences, Engineering and related interdisciplinary fields such as Materials, Nanoscience, Chemical Engineering, Complex Systems and Biophysics.
- The work reported in the thesis must represent a significant scientific advance.
- If the thesis includes previously published material, permission to reproduce this must be gained from the respective copyright holder (a maximum 30% of the thesis should be a verbatim reproduction from the author’s previous publications).
- They must have been examined and passed during the 12 months prior to nomination.
- Each thesis should include a foreword by the supervisor outlining the significance of its content.
- The theses should have a clearly defined structure including an introduction accessible to new PhD students and scientists not expert in the relevant field.

Indexed by zbMATH.

Ryosuke Sei

Two-Dimensional Superconductivity in Rare Earth Oxybismuthides with Unusual Valent Bismuth Square Net

Doctoral Thesis accepted by
The University of Tokyo, Tokyo, Japan

Springer

Author
Dr. Ryosuke Sei
AGC Inc.
Tokyo, Japan

Supervisor
Prof. Tetsuya Hasegawa
The University of Tokyo
Tokyo, Japan

ISSN 2190-5053 ISSN 2190-5061 (electronic)
Springer Theses
ISBN 978-981-99-7312-5 ISBN 978-981-99-7313-2 (eBook)
https://doi.org/10.1007/978-981-99-7313-2

This Springer imprint is published by the registered company Springer Nature Singapore Pte Ltd.
The registered company address is: 152 Beach Road, #21-01/04 Gateway East, Singapore 189721, Singapore

Supervisor's Foreword

The first research goal of the author was the thin film epitaxy of Y_2O_2Bi with a *negatively divalent* Bi square net. The related bulk polycrystalline compounds R_2O_2Bi (R = rare earth) have been just found to show metal insulator transition by Prof. Hideo Hosono's group. The author tried the thin film epitaxy by using the pulsed laser deposition method in ultrahigh vacuum to obtain the negatively divalent Bi. But it was very difficult because he always obtained Y_2O_3 doped with positively trivalent Bi, which was a very stable valence. Then, the author developed reductive solid-phase epitaxy, which enabled to reduce Bi powder by alloying with Y powder at first heating followed by the solid-phase epitaxy of Y_2O_2Bi on single crystal substrate at second heating. But the crystallinity of the film was not satisfactory. Finally, he invented multilayer solid-phase epitaxy by using sputtering. This method is in situ annealing at high temperature, resulting in significantly improved crystallinity. These research were successfully completed, but were just the beginning of other emerging fields. He found a trace of superconductivity in the highly crystalline Y_2O_2Bi epitaxial thin film although the bulk polycrystal was not reported to be superconducting in the previous study. Then he switched his research from thin film growth to bulk crystal growth. He finally found that the excess oxygen nonstoichiometry was a key to the superconductivity in Y_2O_2Bi. Then, he synthesized other R_2O_2Bi. Some of them were hard to oxidize, but he developed a new efficient oxidant CaO. Eventually, he made various R_2O_2Bi superconducting. Surprisingly, the superconducting transition temperatures of R_2O_2Bi were universally scaled by tetragonality, c/a (a, c: lattice constants). The tetragonal distortion of the unit cell governs the superconductivity, regardless of the antiferromagnetic ground state in several R_2O_2Bi.

His achievements opened new fields. The multilayer solid-phase epitaxy is also useful for the first thin film epitaxy of various kinds of compounds, e.g., iso-structural R_2O_2Sb with negatively divalent Sb square net and other pnictides. The tetragonality induced superconductivity in R_2O_2Bi is universally observed not only by oxygen intercalation but also by H or F substitution and Li intercalation. His hard work with his deep intellectual pursuits and without borders of experimental techniques generated such fruitful research.

September 2023 Tetsuya Hasegawa

Parts of this thesis have been published in the following journal articles

1. R. Sei, T. Fukumura, T. Hasegawa, *Crystal Growth and Design* **14**, 4227 (2014).
2. R. Sei, T. Fukumura, T. Hasegawa, *ACS Applied Materials and Interfaces* **7**, 24998 (2015).
3. R. Sei, S. Kitani, T. Fukumura, H. Kawaji, T. Hasegawa, *Journals of the American Chemical Society* **138**, 11085 (2016).
4. H. Kawasoko, K. Ohoyama, R. Sei, K. Matsumoto, D. Oka, A. Hoshikawa, T. Ishigaki, T. Fukumura, *AIP Advances* **9**, 115301 (2019).
5. R. Sei, H. Kawasoko, K. Matsumoto, M. Arimitsu, K. Terakado, D. Oka, S. Fukuda, N. Kimura, H. Kasai, E. Nishibori, K. Ohoyama, A. Hoshikawa, T. Ishigaki, T. Hasegawa, T. Fukumura, *Dalton Transaction* **49**, 3321 (2020).
6. H. Kawasoko, T. Fukumura. *iScience* **25**, 104742 (2022).

Acknowledgements

First of all, I would like to show my greatest appreciation to my supervisor Prof. Dr. Tetsuya Hasegawa who provided me with a wonderful experimental environment and incisive comments. In addition, he always encouraged me through the daily discussions and by taking me for a drink. Those experiences were a very pleasant time for me. My study during the Ph.D. course was successfully done, thanks to his tremendous support.

My deepest appreciation goes to Prof. Dr. Tomoteru Fukumura. His insightful advise and constructive suggestion always helped me to conduct experiments. Fruitful discussions with him motivated me and gave me deep knowledge. Furthermore, he also gave me the great opportunity, not only for my academic life in Sendai but also for the enjoyable moments over a beer. These memories are truly irreplaceable.

I would like to express special thanks to Dr. Taniyuki Furuyama, Dr. Daichi Oka, Dr. Hideyuki Kawasoko, and Ms. Madoka Suzuki, who are/were the staffs in the Fukumura Group in Tohoku University. Thanks to them, I successfully completed my study and had a very fruitful time in Sendai.

I also thank Dr. Yasushi Hirose, Dr. Akira Chikamatsu, Dr. Hideyuki Kamisaka, Mr. Shoichiro Nakao, Dr. Kei Shigematsu, Mrs. Mie Umino, and Mrs. Miki Komazawa for their useful comments and warm encouragement, who are/were the staffs in the Hasegawa Group.

Special thanks to Dr. Daisuke Ogawa, Dr. Jie Wei, Dr. Shungo Kojima, Dr. Thantip S. Krasienapibal, Mr. Shun Inoue, Mr. Kenichi Kaminaga, Mr. Kyouhei Yamatake, Mr. Yutaka Uchida, Mr. Dai Kutsuzawa, Mr. Sho Shitanda, Mr. Yuichi Takahashi, Mr. Takuma Takeda, Mr. Shunsuke Shibata, Mr. Shintaro Fukuda, Ms. Fatima Zainab, Mr. Yuta Koshimoto, Mr. Daichi Saito, Mr. Yuki Saba, Mr. Kyohei Terakado, Mr. Noriyuki Ishigane, Mr. Hiroki Senmaru, Mr. Masaya Honda, Mr. Kota Matsumoto, Mr. Taku Yamamoto, and Mr. Subaru Yusa for their support in experiments and daily discussions, who are/were the members of the Fukumura Group in both the University of Tokyo and Tohoku University. Especially, Mr. Takeda, Mr. Shibata, Mr. Terakado, Mr. Ishigane, Mr. Honda, and Mr. Matsumoto worked together with me for the study about $R2O2Bi$ compounds. I was very pleased with their help. In addition, it is a very good memory to work with drinking beer and talking about baseball at Hongo almost every night with Dr. Ogawa, Dr. Kojima, and Mr. Shibata.

I would like to express my gratitude to Ms. Mayuko Oka, Ms. Kaori Kurita, and Mr. Ryosuke Takagi who are my contemporaries at the Hasegawa Group. Although our studies were not so related to each other, I was very impressed when all of them gathered at Prof. Fukumura's celebration. I also thank other members of Hasegawa Group in the University of Tokyo for their persistent help.

I am deeply grateful to Prof. Dr. Hitoshi Kawaji and Dr. Suguru Kitani for their kind cooperation in conducting specific heat measurement. Thanks to them, my study has been a wonderful achievement.

I am also grateful to Prof. Dr. Masahiro Yamashita and Dr. Shinya Takaishi who lent me a glovebox for my study in Tohoku University. Thanks to them, I was able to resume the experiment about bulk polycrystalline samples from a relatively early time in Sendai.

A part of specific measurements was performed by using a facility of the Cryogenic Research Center, the University of Tokyo.

XPS measurements were performed by using a facility of Nano-Engineering Research Center, Institute of Engineering innovation, School of Engineering, the University of Tokyo.

The low-temperature X-ray diffraction measurement was performed by using a facility of the Institute for Solid State Physics, the University of Tokyo. I also thank Dr. Takeshi Yajima for his kind help to conduct the measurement.

A part of magnetic measurements was performed by using a facility of the Center for Low Temperature Science, Tohoku University.

I would like to give special thanks to the Advanced Leading Graduate Course for Photon Sciences (ALPS) program for providing the scholarship. I also thank Prof. Takeaki Ozawa who is my secondary supervisor in the ALPS program for his encouragement.

Finally, I am deeply grateful to my family and friends for their understanding, moral support, and warm encouragement. Thank you very much.

Contents

Chapter 1
General Introduction

1.1 Layered Superconductors

Since the discovery of high-temperature superconductivity in cuprates [1], naturally and artificially layered compounds have attracted great interest to induce fascinating superconductivity with high transition temperature and unconventional pairing mechanisms. Extensive studies have developed this field deeper and wider, resulting in the discovery of a wide variety of novel layered superconductors such as ruthenates, nitride chlorides, $LaAlO_3/SrTiO_3$ heterostructures, and Fe-based superconductors [2–6]. In this section, I introduce the properties of layered superconductors, especially the cuprate and Fe-based high-temperature superconductors.

1.1.1 High-Temperature Superconductors

The most famous and intriguing nature in layered superconductors is the high transition temperature observed in cuprates and Fe-based compounds. These compounds are composed of two-dimensional building blocks such as superconducting layer and blocking layer. The latter has large flexibility of crystal structures and constituent elements, enabling to engineer the various materials and transition temperatures.

Figure 1.1 shows the crystal structures of mother compounds of well-known cuprate superconductors: La_2CuO_4, $YBa_2Cu_3O_7$, and $Bi_2Sr_2CaCu_2O_8$. All compounds possess two-dimensional square CuO_2 sheet showing superconductivity with various blocking layers. La_2CuO_4 was a first series of high-temperature superconductors reported in 1986 [1] and T_c rose to around 36 K in $La_{1.8}Sr_{0.2}CuO_4$ [7]. $YBa_2Cu_3O_7$ is the first superconductor possessing higher T_c of 93 K than the boiling point of liquid N_2 [8]. Bismuth–strontium–calcium–copper oxide compounds can be represented by the chemical formula $Bi_2Sr_2Ca_{n-1}Cu_nO_{2n+4}$ [9]. These compounds have large blocking layers and T_c increased over 100 K for $n = 3$. These supercon-

R. Sei, *Two-Dimensional Superconductivity in Rare Earth Oxybismuthides with Unusual Valent Bismuth Square Net*, Springer Theses,
https://doi.org/10.1007/978-981-99-7313-2_1

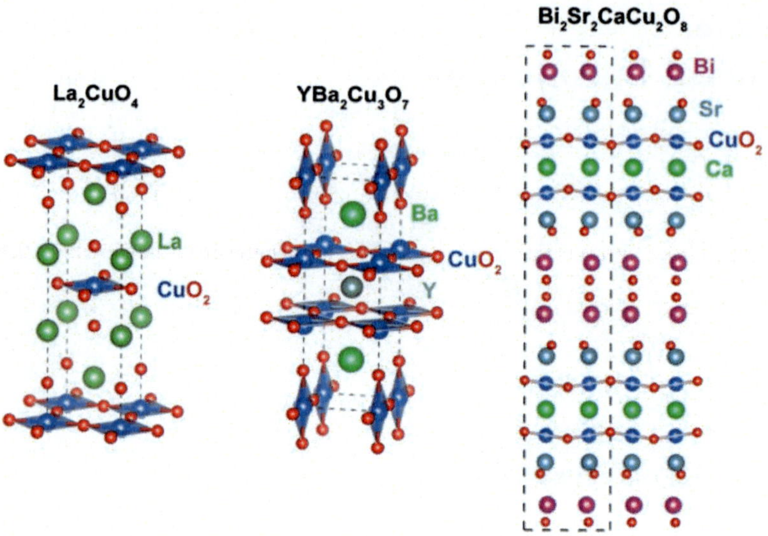

Fig. 1.1 Crystal structures of mother compounds of cuprate superconductors. Broken lines denote unit cells. In this thesis, the crystal structures were drawn by VESTA [10]

ductors are the promising materials for the future application to transmission cables without energy loss, superconducting maglev, and so on.

Another family of high-temperature superconductor is the Fe-based compounds (Fig. 1.2). The first reported compound is $LaFeAsO_{1-x}F_x$ with T_c of 26 K [6]. In this compound, PbO-type FeAs layer composed of edge sharing $FeAs_4$ tetrahedron becomes superconducting, and T_c increases over 50 K by the substitution of rare earth elements [11]. The Fe-based superconductors also form a wide group of materials by converting the crystal structure of a blocking layer such as $BaFe_2As_2$ (T_c =38 K) [12] and LiFeAs (T_c =18 K) [13]. Furthermore, iso-structural layered iron chalcogenides like FeSe also become superconducting (T_c =8 K) [14].

1.1.2 Two-Dimensional Superconductivity

Layered superconductors own two-dimensional electronic nature. This leads to the novel properties, e.g., different upper critical fields between parallel and perpendicular directions of applied magnetic fields to superconducting layer [2].

One characteristic phenomenon in a two-dimensional superconductor is the Berezinskii–Kosterlitz–Thouless (BKT) transition, clearly observed at the superconducting oxide interface in $LaAlO_3/SrTiO_3$ heterostructure [5]. In the two-dimensional superconducting system, vortex–antivortex pairs form the ionic-like lattice, and melt at characteristic BKT transition temperature (T_{BKT}). This behavior appears as $V \propto I^{\alpha}$

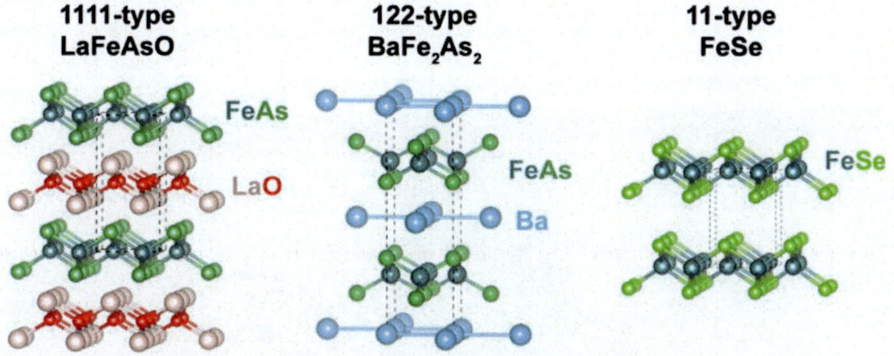

Fig. 1.2 Crystal structures of mother compounds of Fe-based superconductors. Broken lines denote unit cells

law in voltage–current characteristics where α turns from 1 (ohmic law) above T_c to 3 at T_{BKT} and more by decreasing the temperature [15, 16]. The T_{BKT} is independently derived from electrical resistivity [17]. These separately derived temperatures show the same value, proving the BKT transition. This transition is also observed in both cuprate and Fe-based superconductors [18–21], confirming the two-dimensional electronic nature and weak inter-layer interaction of layered superconductors.

1.1.3 Complex Phase Diagram

Figure 1.3a shows a schematic image of a typical superconducting phase diagram in cuprate superconductors [2]. The mother compounds are antiferromagnetic (AF) Mott insulator. By introducing carriers into the CuO_2 plane, the long-range AF ordering is weakened and breaks resulting in the emergence of superconductivity. It is considered that this AF interaction is the driving force of Cooper pairing with *d*-wave symmetry in stark contrast to that by electron–phonon coupling in BCS-type *s*-wave superconductors [22]. However, recent studies report the competition between superconductivity and charge density wave order indicating the presence of strong electron–phonon coupling [23, 24]. These results complicate the elucidation of the superconducting mechanism of cuprates. Fe-based superconductors also show the complex phase competition among superconductivity, antiferromagnetism with spin density wave, and structural transition, although the detailed superconducting mechanism is still unclear [25, 26].

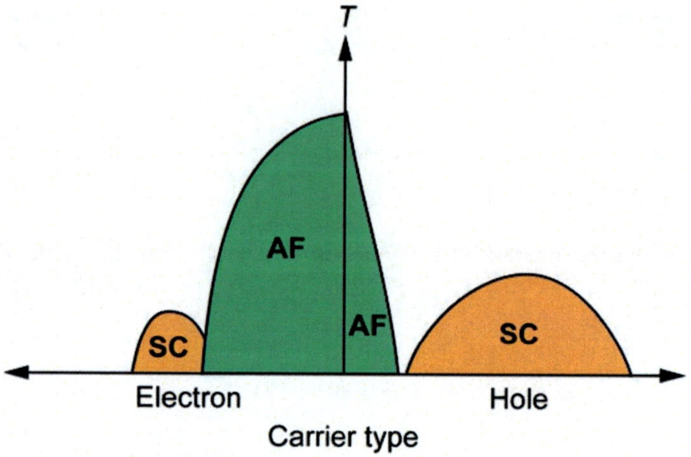

Fig. 1.3 Schematic images of superconducting phase diagram for cuprates. SC and AF denote superconducting and antiferromagnetism, respectively [2]

1.1.4 Schemes to Search Novel Layered Superconductors

There are few layered compounds which show the superconductivity as their stoichiometric composition and at ambient pressure, so carrier doping and applying pressure are widely utilized to induce superconductivity [2]. The carrier doping techniques mainly have two approaches: alio-valent ion substitution and ionic intercalation into van del Waals gap (Fig. 1.4). Generated carriers due to charge neutrality are injected into the conductive layer, changing temperature dependence of electrical resistivity (ρ-T curve) from insulating to superconducting behavior in $La_{2-x}Sr_xCuO_4$ (Fig. 1.5 left) [27]. The high-pressure environment is generally realized by the diamond anvil cell up to over 700 GPa maximum [28], also resulting in the emergence of superconductivity (Fig. 1.5 right) [29].

Exploring iso-structural compounds of layered superconductors is one promising strategy to discover novel superconductors. For example, Sr_2RuO_4 has the same structure as La_2CuO_4 with quite low T_c of around 1.5 K. This compound is considered to possess spin-triplet Cooper pairing with p-wave symmetry, which will induce exotic quantum phenomena [3]. Furthermore, the exploration of the layered compounds structurally similar to Fe-based superconductors develops the generation of a large variety of superconducting families such as layered Ti_2O-based and BiS_2-based superconductors [26].

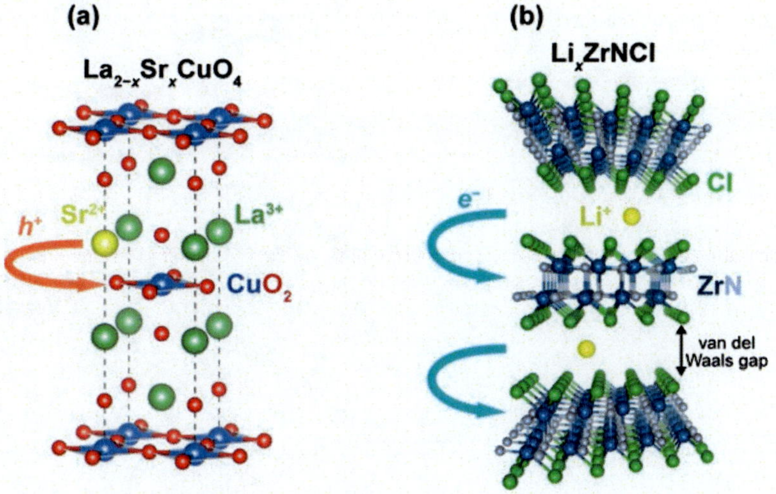

Fig. 1.4 Schematic images of carrier doping techniques: **a** alio-valent ion substitution and **b** ionic intercalation into van del Waals gap

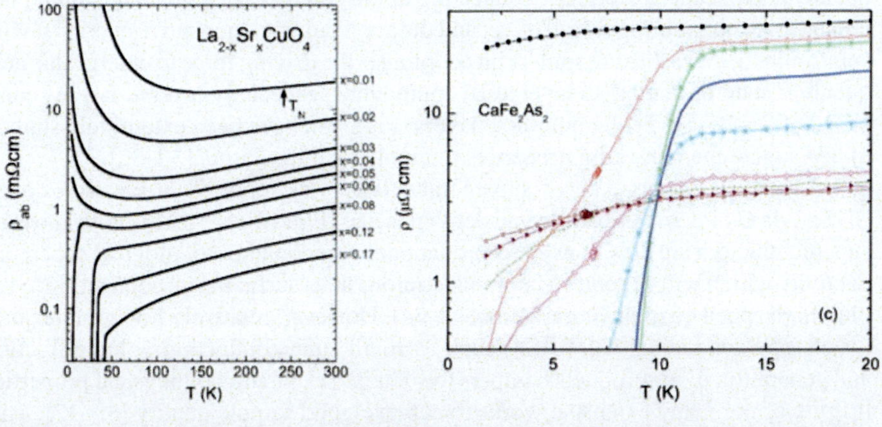

Fig. 1.5 **a** The ρ-T curves of $La_{2-x}Sr_xCuO_4$ with various x (left). Reprinted with permission from [27]. Copyright 2001 by American Physical Society. **b** The ρ-T curves of $CaFe_2As_2$ under pressure at 0, 2.3, 3.5, 5.1, 5.5, 8.6, 12.7, 16.8, and 19.3 kbar. Reprinted with permission from [29]. Copyright 2008 by American Physical Society

1.1.5 Superconductivity in Strong Spin–Orbit Coupled System

In the field of condensed matter physics, topological insulator has attracted much attention for these ten years, which is a bulk insulator but possessing conductive edge/surface state with linear band dispersion called a Dirac cone [30]. The completely spin-polarized massless carriers can propagate at this edge/surface state with-

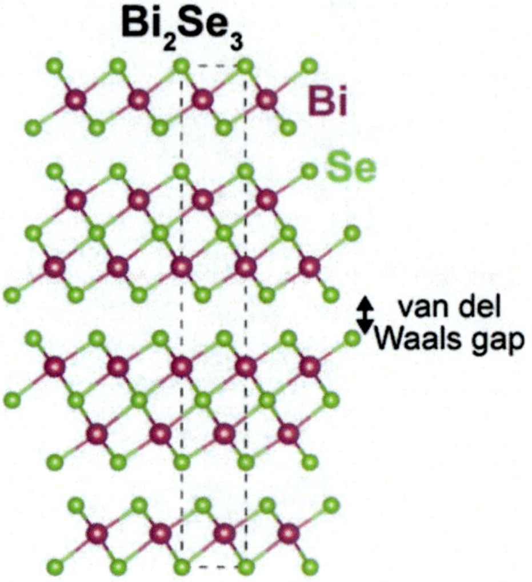

Fig. 1.6 A crystal structures of Bi_2Se_3. Broken lines denote unit cells

out any back-scattering effects, connecting to the emergence of novel quantum phenomena like quantum spin Hall effect and the application to innovative energy-saving spintronics devices. Strong spin–orbit coupling is the driving force to emerge this new quantum state of matter, so layered Bi compounds, especially layered Bi_2Se_3 composed of Se–Bi–Se–Bi–Se quintuple layers (Fig. 1.6), have been extensively studied to investigate the intriguing properties in topological insulators [30–32].

In 2010, the emergence of superconductivity was reported in electron doped Bi_2Se_3 via Cu intercalation into van del Waals gap (Fig. 1.6) [33]. This novel layered superconductors are now considered as a model compound of topological superconductivity which will generate Majorana fermions at its surface state connecting to the ultra high speed quantum computing [31, 34]. However, relatively low sample quality of superconducting $Cu_xBi_2Se_3$ such as small superconducting volume fraction and generation of by-products hampers the detailed evaluation of physical properties in spite of the many extensive works to achieve good sample quality [33–37]. This indicates that the searching novel candidates of topological superconductors is quite a meaningful issue to investigate the interaction between (quasi-)two-dimensional superconductivity and strong spin–orbit coupling in both fundamental and practical aspects.

1.2 Bi Square Net Compounds

Bi is a heavy semimetal element whose atomic number is 83 possessing the strong spin–orbit coupling. The electronic configuration of Bi is $[Xe]4f^{14}5d^{10}6s^26p^3$, so Bi generally possesses +5, +3, and −3 valent states with various local structures

Table 1.1 Crystal structure of $AM\mathrm{Bi}_2$ compounds with Bi^- square net (R: rare earth)

Structure	Compounds
$ZrCuSi_2$ (*P4/nmm*)	$CaMnBi_2$ [43], $YbMnBi_2$ [44], $LaLiBi_2$ [45], $R\mathrm{NiBi}_{-2}$ [46], $R\mathrm{PdBi}_2$ [47], $R\mathrm{CuBi}_2$ [48], $R\mathrm{AgBi}_2$ [49], $R\mathrm{AuBi}_2$ [50], $R\mathrm{ZnBi}_2$ [51]
$SrZnBi_2$ (*I4/mmm*)	$SrMnBi_2$, $BaMnBi_2$ [52], $EuMnBi_2$ [53], $CeGeBi_2$ [40], $BaZnBi_2$, $SrCdBi_2$, $BaCdBi_2$ [54]

in compounds. For example, elemental Bi forms a buckled honeycomb lattice [38], and Bi^{3+} and Bi^{5+} form octahedra with anions such as BiO_6 in $BaBiO_3$ [39] and $BiSe_6$ in Bi_2Se_3 [33], respectively. Bi^{3-} anions can be realized by the coexistence of highly electropositive elements such as alkaline earth and rare earth, forming the complicated polyhedra with these cations [40]. Notably, Bi can form the monatomic square net structure with an incomplete filled electron shell such as Bi^- state and unusually a reductive Bi^{2-} state. This section reviews the fascinating properties of Bi square net compounds.

1.2.1 Crystal Structure

There are mainly two types of Bi square net compounds whose chemical formulas are represented by $AM\mathrm{Bi}_2$ (A: alkaline earth and rare earth; M: metal element) and $R_2X_2\mathrm{Bi}$ (R: rare earth and actinoid; X: O and N) [40, 41]. The former possesses a Bi^- square net in stark contrast to an unusual Bi^{2-} square net in the latter.

There is a large variety of $AM\mathrm{Bi}_2$ compounds because of the flexibility of the combination between A and M. Figure 1.7 shows crystal structures of $AM\mathrm{Bi}_2$ compounds which can be further classified into two structures: $ZrCuSi_2$-type (space group: *P4/nmm*) and $SrZnBi_2$-type (*I4/mmm*) [40, 42]. In both structures, the Bi^- square net is separated from the edge sharing $M\mathrm{Bi}_4$ tetrahedra slab (same as PbO-type structure) by the interstitial A metals, and the classification is originating from the different arrangement of A atoms adjacent to Bi^- square net (bottom of Fig. 1.7), resulting in the double-sized unit cell in $SrZnBi_2$-type structure. The compounds' list of these structure types are summarized in Table 1.1. It seems that several compounds such as $R\mathrm{NiBi}_2$ and $R\mathrm{PdBi}_2$ show the intrinsic defect in M site, so that they are often represented as $AM_{1-x}\mathrm{Bi}_2$ [46, 47].

In contrast to a large variation of Bi^- square net compounds, there are a few reports of Bi^{2-} square net compounds: $Ac_2N_2\mathrm{Bi}$ (Ac: Th, U, Am, Cm) [55, 56] and $R_2O_2\mathrm{Bi}$ (R: rare earth) [41, 57–59]. Both groups have an anti-$ThCr_2Si_2$-type structure composed of $[R_2O_2]^{2+}$ (or $[Ac_2N_2]^{2+}$) slab with a PbO-type structure and Bi^{2-} square net (Fig. 1.8a). Here, "anti-" means that the positions of anions and cations are inverted from normal $ThCr_2Si_2$-type compounds, e.g., $BaFe_2As_2$ high-temperature

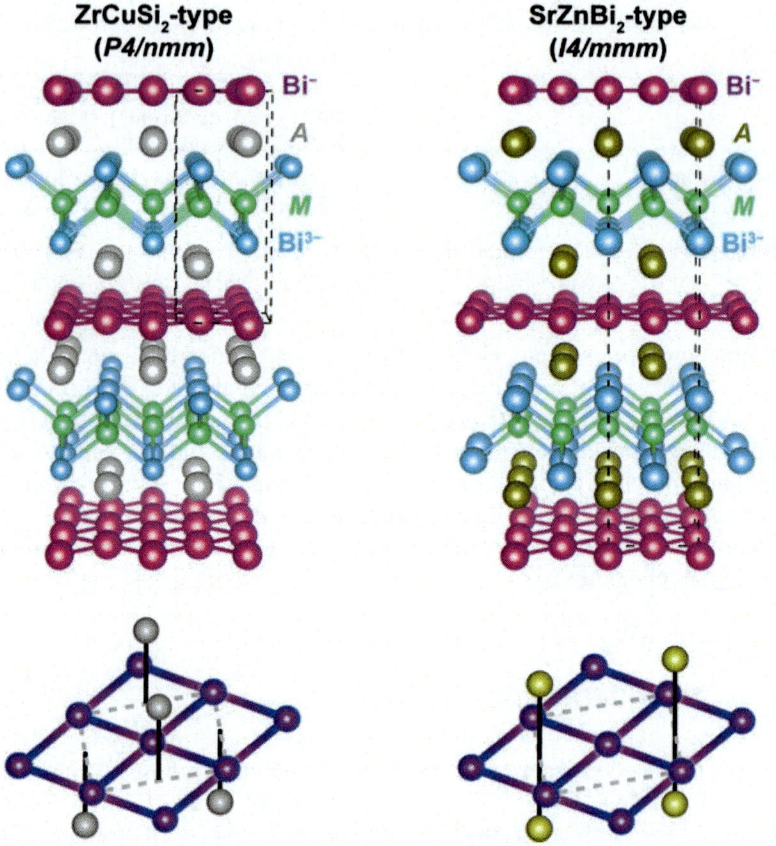

Fig. 1.7 Crystal structures of $ZrCuSi_2$-type (left) and $SrZnBi_2$-type (right) $AMBi_2$ compounds. Broken lines denote unit cells. Bottoms show the local crystal structures around Bi^- square net

superconductor. The space group is *I4/mmm*. The Bi^{2-} square net is rotated by 45° on the PbO-type layer compared with the Bi^- square net in $AMBi_2$. Figure 1.8b shows lattice constants plotted as a function of ionic radii of R^{3+} [60] in R_2O_2Bi reported in previous studies [41, 57–59]. The lattice constants show systematical variation according to the lanthanoid contraction, evidencing the R^{3+} state which indirectly confirms highly reductive Bi^{2-} state. Furthermore, larger Bi–Bi bond length in Bi^{2-} square net, i.e., *a*-axis length, than that in Bi^- square net (~3.2 Å) [58] also supports Bi^{2-} state as discussed later.

The multinary compounds composed of electropositive elements (e.g., alkaline earth and rare earth) and electronegative elements (e.g., group 13–16 elements) are called the Zintl phase where the former donates their electrons to the latter [61, 62]. Each electronegative element forms a cluster structure according to the octet rule, so Bi anions generally form the following frameworks: isolated square or zig-zag chain for Bi^-, dimer for Bi^{2-}, and isolated atom for Bi^{3-}. One

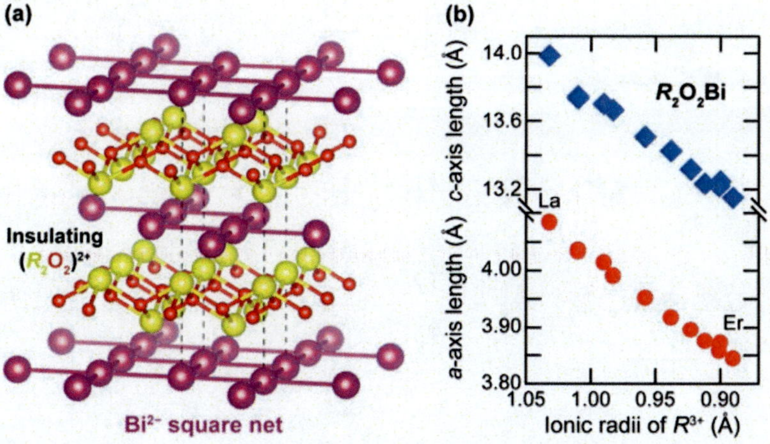

Fig. 1.8 **a** Crystal structure of $ThCr_2Si_2$-type R_2O_2Bi. Broken line denotes unit cell. **b** Lattice constant variation of R_2O_2Bi. Structural data refers to [41, 57–59]

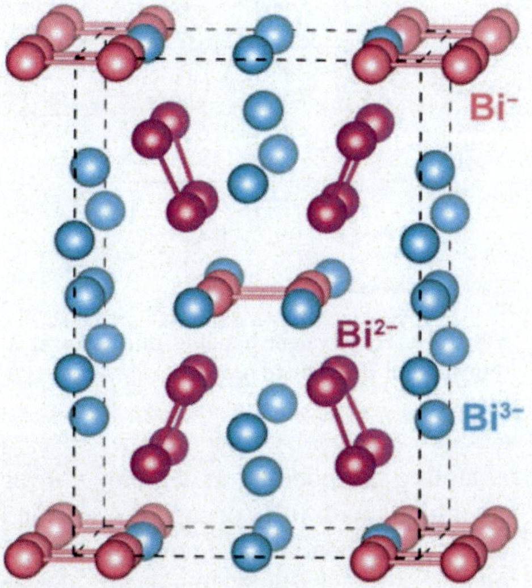

Fig. 1.9 Bi framework in $Ca_{11}Bi_{10}$. Ca cations are not shown here. Broken line denotes unit cell

typical example is $Ca_{11}Bi_{10}$ [63] whose charges of each element are assigned as $(Ca^{2+})_{22}(Bi^{-})4(Bi^{2-})_8(Bi^{3-})_8$, where four Bi^{-} anions form one isolated square, eight Bi^{2-} anions form four dimers, and eight Bi^{3-} anions are isolated (Fig. 1.9). Both *AM*Bi_2 and R_2O_2Bi compounds are considered as a kind of Zintl phase, while a monatomic Bi square net layer cannot be explained by the semiclassical Zintl concept.

These nonclassical structures can be explained by introducing the hypervalent bonding rule [62]. The simplest example is Bi_3^{7-} forming a linear trimer, whose electronic configuration is equal to typical hypervalent product Subscript 3_3 $^{-}$. By

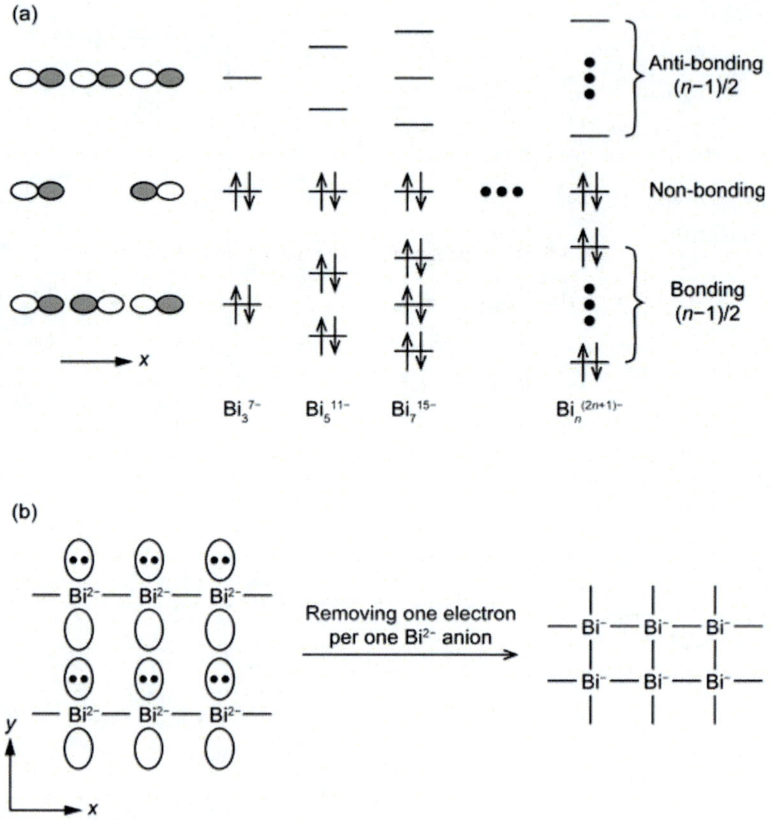

Fig. 1.10 Hypervalent bonding rule to form **a** one-dimensional Bi^{2-} linear chain and **b** two-dimensional Bi^{-} square net [62]

assuming the bonding axis along x direction, there are four electrons in three Bi p_x orbitals with fully occupied s, p_y, and p_z orbitals (Fig. 1.10a). These electrons occupy only bonding and non-bonding molecular orbitals, resulting in the formation of $Bi_3{}^{7-}$ linear trimer. Increase in the number of Bi atoms leads to the formation of $Bi_n^{(2n+1)-}$ linear chain (Fig. 1.10a), converging to one-dimensional Bi^{2-} linear chain structure in the crystal ($n \rightarrow \infty$). The next step is the formation of Bi square net by considering the array of one-dimensional Bi^{2-} linear chain along y direction (Fig. 1.10b). In order to make Bi–Bi bond between filled p_y orbitals of Bi^{2-} anions, it is necessary to remove one electron per one Bi^{2-} anion, resulting in a two-dimensional Bi^{-} square network. Therefore, only Bi^{-} anions can form the square net structure even in a hypervalent bonding rule, indicating unusual condition of Bi^{2-} square net in R_2O_2Bi.

Similar electronic configurations to Bi^{2-} square net are widely observed in chalcogen compounds such as RTe_2 and RTe_3 [62] where Te^{-} anions should form a square net. However, observed square net is quite unstable and incidentally shows Peierls

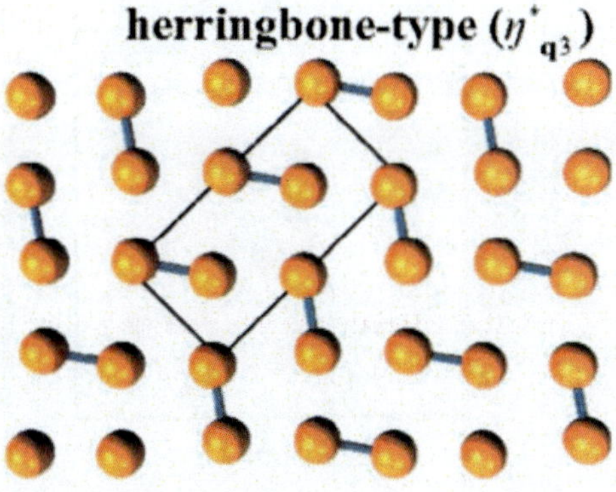

Fig. 1.11 Schematic image of herringbone-type structure. Reprinted with permission from [66]. Copyright 2016 by American Physical Society

distortion, i.e., charge density wave (CDW) transition, because one excess electron occupies an anti-bonding orbital in addition to the favorable dimerization of Te^- due to the octet rule. Indeed, the analogous compounds R_2O_2Sb also show CDW transition where Sb^{2-} anions form dimers resulting in herringbone-type Sb layer (Fig. 1.11) [64, 65]. On the other hand, surprisingly, there are no CDW transitions in R_2O_2Bi compounds [57–59]. Theoretical report indicates that strong spin–orbit coupling and chemical pressure effect suppress CDW instability, unusually retaining the complete Bi^{2-} square net [66]. The longer Bi–Bi bond length in R_2O_2Bi than that in $AMBi_2$ is due to the presence of one anti-bonding electron, evidencing Bi^{2-} state.

1.2.2 *Anisotropic Dirac Fermion in Bi Square Net Compounds*

Although the existence of Bi square net has been known for half a century, the recent discovery of anisotropic Dirac fermion in Bi square net explosively accelerated its research. Figure 1.12 shows angle-resolved photoelectron spectroscopy (ARPES) images of $SrMnBi_2$ [67]. Along Γ-M line (cut 2), the sharp needle-like band dispersion is observed both near Fermi level and at high binding energy, indicating the existence of Dirac fermions. The estimated Fermi velocity of $v_F^{\parallel} \approx 1.1 \times 10^6$ m s^{-1} is comparable to the value of graphene. On the other hand, band dispersion shows curvature near Fermi level perpendicular to Γ-M line (cut 1) with almost linear dispersion at high binding energy resulting in relatively small Fermi velocity of $v_F^{\perp} \sim 2 \times 10^5$ m s^{-1}. This result indicates the large anisotropy of Dirac-like band dispersion with $v_F^{\parallel}/v_F^{\perp} \gtrsim 5$, consistent with the theoretical calculation [42, 67, 68]. This anisotropic Dirac fermion has been only theoretically proposed such as in bent graphene and some nanocomposites [67], probably connecting to the exotic quantum phenomena and application to novel electronic devices with different direction of carrier conduction path. Therefore, naturally formed anisotropic Dirac fermion in $SrMnBi_2$ is valuable and attracts much attention.

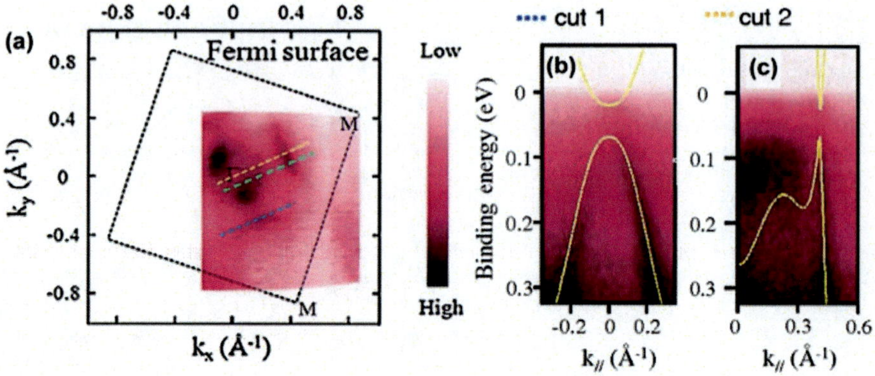

Fig. 1.12 The results of ARPES of $SrMnBi_2$. **a** Fermi surface. Band dispersion **b** perpendicular to Γ-M line (blue broken line in (**a**)) and **c** parallel to Γ-M line (orange broken line in (**a**)). Reprinted with permission from [67]. Copyright 2011 by American Physical Society

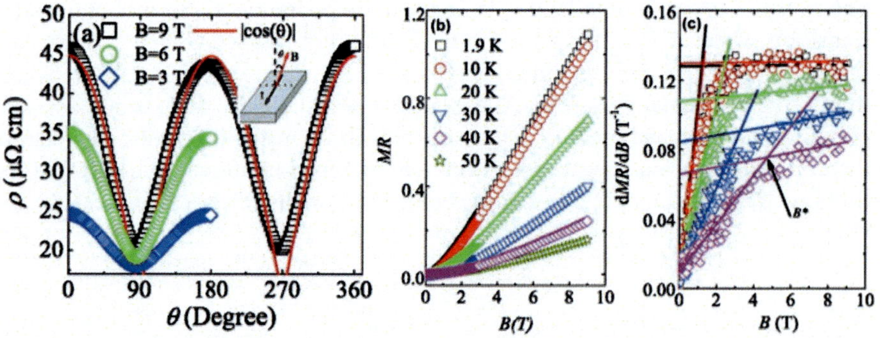

Fig. 1.13 **a** Angular dependence of MR. Inset shows measurement configuration. **b** MR at various temperatures. **c** The field derivative of MR at various temperatures. Reprinted with permission from [69]. Copyright 2011 by American Physical Society

The existence of Dirac fermions in Bi^- square net are also characterized in electrical transport properties [69]. Figure 1.13a shows the angular dependence of magnetoresistance (MR) of $SrMnBi_2$. The MR showed two-fold symmetry with complete fitting by $|\cos\theta|$ function, demonstrating a two-dimensional carrier conductive path, i.e., Bi^- square net. Notably, MR shows unsaturated linear behavior confirmed by constant $\mathrm{d}MR/\mathrm{d}B$ at high field, contrary to a saturation at high field in normal metals (Fig. 1.13b,c). This unsaturated linear MR is a typical phenomenon of linear band dispersive compounds which are experimentally observed and theoretically explained [69, 70]. In addition, small effective mass of carriers ($0.29m_e$) and high carrier mobility ($250\,\mathrm{cm^2\ V\ s^{-1}}$ [67] and $3400\,\mathrm{cm^2\ V\ s^{-1}}$ [69]) are estimated from the MR data. These results strongly support the existence of Dirac fermions in Bi^- square net.

The anisotropic Dirac fermions are also observed in other $AM\mathrm{Bi}_2$ compounds such as $CaMnBi_2$ [71, 72], $LaAgBi_2$ [73], $EuMnBi_2$ [53, 74], $BaMnBi_2$ [75], and $YbMnBi_2$ [44]. It was initially considered that magnetic ordering mediated the formation of anisotropic Dirac fermions because first compounds such as $SrMnBi_2$ and $CaMnBi_2$ show AF ordering of $MnBi_4$. However, a similar behavior in non-magnetic $LaAgBi_2$ suggests the universal existence of anisotropic Dirac fermions in Bi^- square net. Indeed, theoretical studies support this scenario [42, 67, 68]. A single Bi square net can possess Dirac-like band dispersion composed of Bi $6p_x$ and $6p_y$ orbitals. The Dirac band is weakly hybridized with adjacent atoms A in $AM\mathrm{Bi}_2$ compounds especially perpendicular to Γ-M line, resulting in the anisotropy of the Dirac band between parallel and perpendicular to the Γ-M line. Furthermore, there are mainly two effects of A atoms to Dirac band. First, the different arrangement of A atoms adjacent to Bi square net (bottom of Fig. 1.7) causes different hybridization effect demonstrating the flatter band shape of $CaMnBi_2$ than that of $SrMnBi_2$. Second, different kinds of atoms change the strength of the spin–orbit coupling of Bi square net, resulting in the larger bang gap of $SrMnBi_2$ than that of $CaMnBi_2$. These results indicate that the properties of Bi square net can be engineered by designing the surrounding environment of Bi square net.

The AF ordering of $MnBi_4$ in $SrMnBi_2$ and $CaMnBi_2$ seems not to affect the Dirac band formation. However, their different spin ordering states possibly have some influence on electrical transport properties with (without) a kink at T_N in electrical resistivity in $CaMnBi_2$ ($SrMnBi_2$), where $CaMnBi_2$ has ferromagnetic inter-layer coupling and $SrMnBi^2$ has an AF one (Fig. 1.14) [76, 77]. A recent study claims that $YbMnBi_2$ is a first compound possessing magnetic Weyl fermion [78]. In addition, buckled Bi square net is theoretically proposed as a novel two-dimensional topological insulator with a large band gap, connecting to the possible application to a room temperature innovative device [79]. The exploration of Dirac band in Bi square net is still developing, so various exotic phenomena will continue to be induced.

1.2.3 Kondo Lattice Behavior in CeMBi$_2$

Several Ce$M\mathrm{Bi}_2$ compounds such as $CeNiBi_2$ and $CePd_{1-x}Bi_2$ show Kondo lattice behavior originating from Ce^{3+} moment [80, 81]. These compounds show AF ordering of Ce $4f^1$ moment at low temperature ($\sim$ 6 K) with strong anisotropy in the temperature dependence of magnetic susceptibility (χ-T curve). The magnetic easy axes are [001] directions. Notably, the deviation of χ-T curves from the Curie–Weiss law is observed below 50 K for $CeNiBi_2$ and 75 K for $CePd_{1-x}Bi_2$, attributed to the strong effect of a crystalline-electric-field. The interaction between Kondo and a crystalline-electric-field effect causes interesting phenomena in electrical transport properties. Figure 1.15a shows ρ-T curves of $CePd_{1-x}Bi_2$ at various fields. The broad shoulder-like anomaly is observed around 75 K originating from the interaction between Kondo and the crystalline-electric-field effect, which is widely observed in Ce-based compounds [81]. The ρ minimum at 20 K is attributed to the

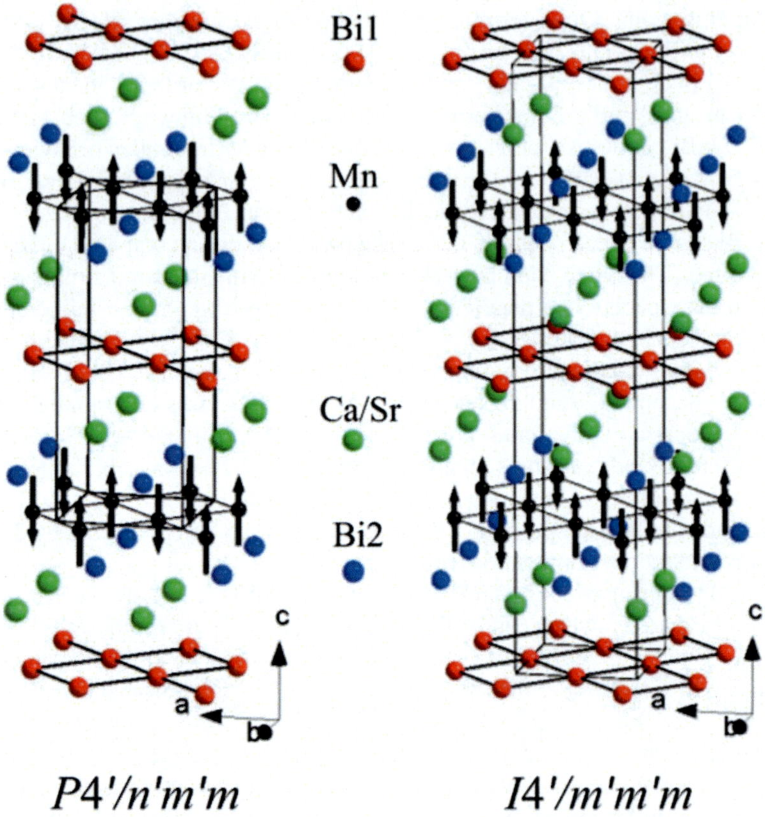

Fig. 1.14 Magnetic structures of $CaMnBi_2$ (left) and $SrMnBi_2$ (right). Reprinted with permission from [76]. Copyright 2014 by American Physical Society

typical Kondo effect. In addition, abrupt ρ drop at 6 K is due to the AF ordering, and this anomaly shifts to lower temperature followed by suppression with increasing magnetic field, because AF ordering connects to metamagnetic transition induced by the field. The MR shows small positive values ($\sim 0.2\%$) at $T > 75K$ in $CePdBi_2$. On the other hand, the MR turns to negative values with larger absolute values at $T < 75$ K (Fig. 1.15b), except for anomaly at $T < T_N$ attributed to the AF to metamagnetic transition. The polarity change of MR is due to the Fermi surface recontraction caused by the interaction between Kondo and the crystalline-electric-field effect around 75 K. The similar complicated interaction is also observed in $CeNiBi_2$ [80].

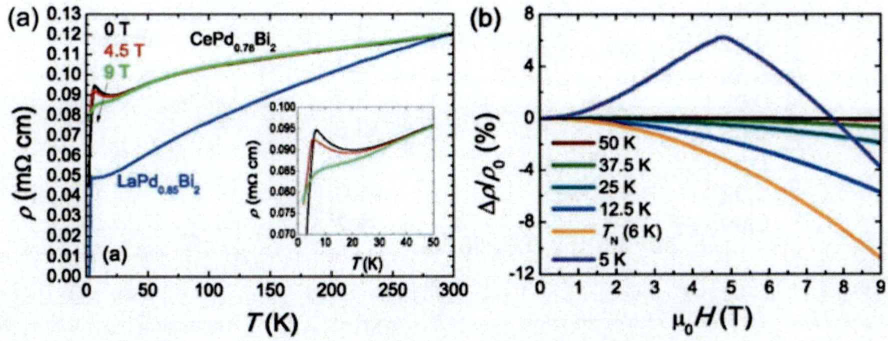

Fig. 1.15 **a** The ρ-T curves of $CePd_{1-x}Bi_2$ at various fields. **b** Magnetoresistance data of $CePd_{1-x}Bi_2$ below 75 K. Reprinted with permission from [81]. Copyright 2015 by American Physical Society

1.2.4 Potential for Superconducting Bi Square Net

The induction of superconductivity in Bi square net is a fascinating issue because it should become an intriguing platform to investigate the interaction between superconductivity and Dirac fermion, probably connecting to the topological superconductivity. In 2011, $CeNi_{1-x}Bi_2$ was reported as a first compound possessing superconducting Bi^- square net with T_c of 4.2 K [82]. The authors claimed that superconductivity associated with the light carriers in Bi^- square net coexisted with the AF ordering at ~5 K of heavy carriers in Ce^{3+}, and Ni deficiency was essential to induce superconductivity. However, this result is not reproduced in subsequent reports [46, 83, 84]: the samples show no Ni deficiency dependence of superconducting properties, no reproducibility of the emergence, and small superconducting volume fraction (Fig. 1.16). Furthermore, T_c of 4.2 K is a similar value to those of NiBi and $NiBi_3$ [85,

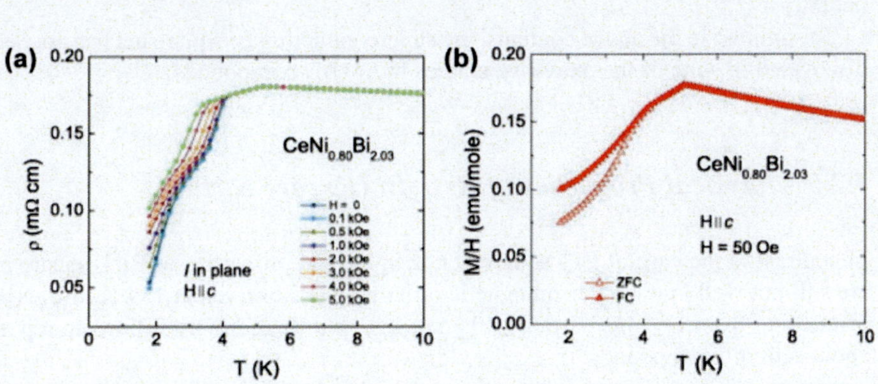

Fig. 1.16 **a** The ρ-T curves at various fields and **b** χ-T curve of $CeNi_{0.8}Bi_2$ single crystal. Reprinted with permission from [46]. Copyright 2013 by Elsevier

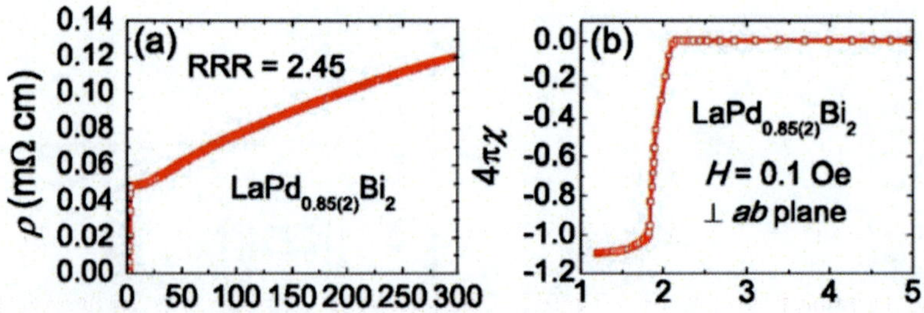

Fig. 1.17 **a** The ρ-T curve and **b** χ-T curve of $LaPd_{0.85(2)}Bi_2$ single crystal. RRR is derived by ρ_{300K}/ρ_{T_c}. Reprinted with permission from [47]. Copyright 2013 by American Physical Society

86], and $LaNi_{1-x}Bi_2$ shows the same T_c and H_{c2} with small superconducting volume fraction [46, 82]. These results indicate that the superconductivity of $CeNi_{1-x}Bi_2$ is not intrinsic but originates from impurity phases such as NiBi and $NiBi_3$.

$LaPd_{1-x}Bi_2$ is reported as superconducting at 2.1 K although rare earth substituted $CePd_{1-x}Bi_2$ shows AF ordering of Ce^{3+} at 6 K without superconductivity as shown in the previous subsection (Fig. 1.17) [47]. Band calculation indicates that Pd orbital has the dominant contribution to form a Fermi surface. In addition, Pd vacancy suppress the Fermi surface nesting, i.e., CDW instability, which seems essential to induce superconductivity. The small residual resistivity ratio (RRR) is attributed to the strong scattering by Pd vacancy, supporting the carrier conduction at Pd layer. Furthermore, the superconductivity is suppressed by compensating Pd vacancy and also the substitution of Fe for Pd, consistent with the band calculation [87]. These results confirm that the superconducting path is the Pd layer.

The superconductivity is also observed at 5.7 K in $SrMnBi_2$ with blob-like Bi residue [88]. However, this superconductivity shows strong sample dependence and is not reproduced. It is considered that the superconductivity is originating from the strained Bi grown on the $SrMnBi_2$ crystals, and not from intrinsic Bi^- square net [88].

To summarize the above contents, the superconducting Bi square net has not been discovered in spite of the extensive studies in $AMBi_2$ compounds.

1.2.5 Physical Properties of R_2O_2Bi (R: rare earth)

In contrast to the case of Bi^- square net, the physical properties of Bi^{2-} square net are still not well known. Ferromagnetic ordering was observed at 154 K in U_2N_2Bi originating from U^{4+} [89] around 40 years ago, but there are no follow-up reports about actinoid compounds.

The promising platform to investigate the intrinsic properties of Bi^{2-} square net is R_2O_2Bi, which was systematically synthesized in polycrystalline powder form in 2011 [58]. Figure 1.18 shows temperature dependence of normalized electrical resis-

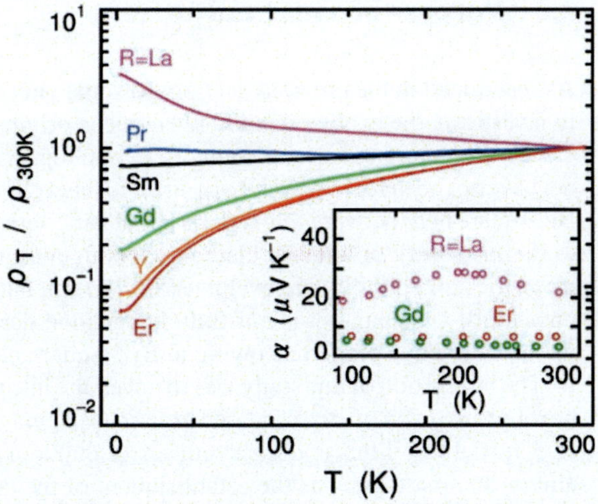

Fig. 1.18 Normalized ρ-T curves of R_2O_2Bi compounds. The inset shows Seebeck coefficients α demonstrating p-type conduction. Reprinted with permission from [58]. Copyright 2011 by American Chemical Society

Table 1.2 Magnetic parameters for R_2O_2Bi compounds. Reprinted with permission from [58]. Copyright 2011 by American Chemical Society

cmpd	T_N (K)	χ_0 (emu mol-R^{-1})	θ (K)	C (emu mol-R^{-1})	P_{eff} (μ_B)
Pr_2O_2Bi	15.0	0.0008	−25.1	1.71	3.70
Gd_2O_2Bi	10.1	0.0011	−25.0	7.80	7.90
Er_2O_2Bi	3.0	0.0021	−19.4	12.32	9.90

tivity $\rho(T)/\rho_{300K}$ in various R_2O_2Bi compounds. Electrical transport property shows insulator-to-metal transition induced by chemical pressure: variation from insulating in La_2O_2Bi to metallic behavior in Er_2O_2Bi. However, the superconductivity is not observed down to 1.8 K [26, 58]. The origin of the insulator-to-metal transition was considered to be attributed to a Mott transition [58], but subsequent theoretical reports suggested that this is due to the stronger electron–phonon coupling resulting in increased electron–phonon scattering in La_2O_2Bi than that in Er_2O_2Bi [66]. The band calculation indicates that the Fermi surface is composed of Bi $6p_x$ and $6p_y$ orbitals of Bi^{2-} square net in R_2O_2Bi, suggesting two-dimensional electronic nature [58, 66]. The χ-T curves show a sharp drop around 10 K in magnetic R_2O_2Bi (R: Pr, Gd, Er) indicating the AF ordering of $4f$ spins in R cations [58]. The χ-T curves are well fitted by the Curie–Weiss law $\chi = \chi_0 + C/(T - \theta)$ at $T > T_N$, demonstrating similar effective Bohr magneton values P_{eff} to theoretical values of R^{3+} state (Table 1.2), evidencing unusual Bi^{2-} state.

1.3 Purpose of This Study

As mentioned in the previous section, Bi square net compounds are a good platform to investigate the exotic quantum phenomena originating from the two-dimensional electronic nature and strong spin–orbit coupling, while superconducting Bi square net have not been realized yet. In contrast to the well-studied $AMBi_2$ compounds with Bi_- square net, there are few reports about Bi^{2-} square net compounds probably due to the instability of unusual electronic configuration. The electrical resistivity and magnetic susceptibility are evaluated in R_2O_2Bi, but detailed physical properties of unusual Bi^{2-} square net is not fully understood despite the possible emergence of various attractive properties owing to Bi^{2-} square net.

The motivation of this study was to unveil the intrinsic properties and to induce the novel phenomena of Bi^{2-} square net in R_2O_2Bi via two approaches: the fabrication of epitaxial thin films and the synthesis of bulk polycrystalline powders. The main aim of the first topic was the establishment of the universal fabrication method for R_2O_2Bi epitaxial thin films which are ideal samples to evaluate the intrinsic physical properties. I succeeded in the achievement of the epitaxial thin films by developing two novel fabrication techniques utilizing solid-phase epitaxy. In addition, the influence of two-dimensional electronic nature and the strong spin–orbit coupling on electrical transport properties was investigated for the first time. The purpose of the second topic was the control of physical properties of Bi^{2-} square net by tuning the composition and the crystal structure of R_2O_2Bi in the form of polycrystalline powders, which are relatively easy to synthesize in comparison with epitaxial thin film. I aimed to construct a guideline for properties engineering of Bi^{2-} square net compounds as a counterpart of various Bi_- square net compounds described above, resulting in the discovery of the universal two-dimensional superconductivity of Bi^{2-} square net.

References

1. Bednorz JG, Müller KA (1986) Zeitschrift für Physik B Condensed Matter 64(2):189
2. Klemm RA (2012) Layered superconductors, vol 1. Oxford University Press
3. Mackenzie AP, Maeno Y (2003) Rev Mod Phys 75:657
4. Yamanaka S, Hotehama KI, Kawaji H (1998) Nature 392(6676):580
5. Reyren N, Thiel S, Caviglia A, Kourkoutis LF, Hammerl G, Richter C, Schneider C, Kopp T, Rüetschi AS, Jaccard D et al (2007) Science 317(5842):1196
6. Kamihara Y, Watanabe T, Hirano M, Hosono H (2008) J Am Chem Soc 130(11):3296
7. Cava RJ, van Dover RB, Batlogg B, Rietman EA (1987) Phys Rev Lett 58:408
8. Wu MK, Ashburn JR, Torng CJ, Hor PH, Meng RL, Gao L, Huang ZJ, Wang YQ, Chu CW (1987) Phys Rev Lett 58:908
9. Maeda H, Tanaka Y, Fukutomi M, Asano T (1988) Jpn J Appl Phys 27(2A):L209
10. Momma K, Izumi F (2011) J App Crystallogr 44:1272
11. Zhi-An R, Wei L, Jie Y, Wei Y, Xiao-Li S, Guang-Can C, Xiao-Li D, Li-Ling S, Fang Z, Zhong-Xian Z et al (2008) Chin Phys Lett 25(6):2215
12. Rotter M, Tegel M, Johrendt D (2008) Phys Rev Lett 101:107006

13. Wang X, Liu Q, Lv Y, Gao W, Yang L, Yu R, Li F, Jin C (2008) Solid State Commun 148(11):538
14. Hsu FC, Luo JY, Yeh KW, Chen TK, Huang TW, Wu PM, Lee YC, Huang YL, Chu YY, Yan DC et al (2008) Proc Natl Acad Sci 105(38):14262
15. Berezinskii V (1971) Soviet J Exp Theoret Phys 32:493
16. Kosterlitz JM, Thouless DJ (1973) J Phys C Solid State Phys 6(7):1181
17. Halperin B, Nelson DR (1979) J Low Temp Phys 36(5–6):599
18. Yeh NC, Tsuei CC (1989) Phys Rev B 39:9708
19. Li Q, Hücker M, Gu GD, Tsvelik AM, Tranquada JM (2007) Phys Rev Lett 99:067001
20. Schneider R, Zaitsev A, Fuchs D et al (2014) J Phys Condens Matter 26(45):455701
21. Lin Z, Mei C, Wei L, Sun Z, Wu S, Huang H, Zhang S, Liu S, Feng Y, Tian H et al (2015) Sci Rep 5
22. Tinkham H (2004) Introduction to superconductivity, 2nd ed. Dover Publication, Inc
23. Ghiringhelli G, Le Tacon M, Minola M, Blanco-Canosa S, Mazzoli C, Brookes N, De Luca G, Frano A, Hawthorn D, He F et al (2012) Science:1223532
24. Chang J, Blackburn E, Holmes A, Christensen N, Larsen J, Mesot J, Liang R, Bonn D, Hardy W, Watenphul A et al (2012) Nat Phys 8(12):871
25. Dong J, Zhang H, Xu G, Li Z, Li G, Hu W, Wu D, Chen G, Dai X, Luo J et al (2008) EPL (Europhys Lett) 83(2):27006
26. Hosono H, Tanabe K, Takayama-Muromachi E, Kageyama H, Yamanaka S, Kumakura H, Nohara M, Hiramatsu H, Fujitsu S (2016) Sci Technol Adv Mater
27. Ando Y, Lavrov AN, Komiya S, Segawa K, Sun XF (2001) Phys Rev Lett 87:017001
28. Dubrovinsky L, Dubrovinskaia N, Bykova E, Bykov M, Prakapenka V, Prescher C, Glazyrin K, Liermann HP, Hanfland M, Ekholm M et al (2015) Nature 525(7568):226
29. Torikachvili MS, Bud'ko SL, Ni N, Canfield PC (2008) Phys Rev Lett 101:057006
30. Hasan MZ, Kane CL (2010) Rev Mod Phys 82:3045
31. Ando Y (2013) J Phys Soc Jpn 82(10):102001
32. Isaeva S, Rasche B, Ruck M (2013) Physica Status Solidi (RRL)-Rapid Res Lett 7(1–2):39
33. Hor Y, Williams A, Checkelsky J, Roushan P, Seo J, Xu Q, Zandbergen H, Yazdani A, Ong N, Cava R (2010) Phys Rev Lett 104(5):057001
34. Sasaki S, Kriener M, Segawa K, Yada K, Tanaka Y, Sato M, Ando Y (2011) Phys Rev Lett 107(21):217001
35. Wang YL, Xu Y, Jiang YP, Liu JW, Chang CZ, Chen M, Li Z, Song CL, Wang LL, He K, Chen X, Duan WH, Xue QK, Ma XC (2011) Phys Rev B 84:075335
36. Shirasawa T, Sugiki M, Hirahara T, Aitani M, Shirai T, Hasegawa S, Takahashi T (2014) Phys Rev B 89:195311
37. Liu Z, Yao X, Shao J, Zuo M, Pi L, Tan S, Zhang C, Zhang Y (2015) J Am Chem Soc 137(33):10512
38. Cucka P, Barrett C (1962) Acta Crystallogr A 15(9):865
39. Cox D, Sleight A (1976) Solid State Commun 19(10):969
40. Gschneidner K, Bünzli J, Pecharsky V (2006) Handbook on the physics and chemistry of rare earths, chap. 227, vol 36. Elsevier Science. https://books.google.co.jp/books?id=_Ro3Fqtz4xgC
41. Benz R (1971) Acta Crystallogr Sect B Struct Crystallogr Cryst Chem 27(4):853
42. Lee G, Farhan MA, Kim JS, Shim JH (2013) Phys Rev B 87(24):245104
43. Brechtel E, Cordier G, Schäfer H (1980) Zeitschrift für Naturforschung B 35(1):1
44. Wang A, Zaliznyak I, Ren W, Wu L, Graf D, Garlea VO, Warren JB, Bozin E, Zhu Y, Petrovic C (2016) Phys Rev B 94:165161
45. Pan DC, Sun ZM, Mao JG (2006) J Solid State Chem 179(4):1016
46. Lin X, Straszheim WE, Bud'ko SL, Canfield PC (2013) J Alloy Compd 554:304
47. Han F, Malliakas CD, Stoumpos CC, Sturza M, Claus H, Chung DY, Kanatzidis MG (2013) Phys Rev B 88(14):144511
48. Jesus C, Piva M, Rosa P, Adriano C, Pagliuso P (2014) J Appl Phys 115(17):17E115
49. Petrovic C, Bud'ko S, Strand J, Canfield P (2003) J Magn Magn Mater 261(1):210
50. Seibel EM, Xie W, Gibson QD, Cava R (2015) J Solid State Chem 230:318

51. Zelinska OY, Mar A (2008) J Alloy Compd 451(1):606
52. Cordier G, Schäfer H (1977) Zeitschrift für Naturforschung B 32(4):383
53. May AF, McGuire MA, Sales BC (2014) Phys Rev B 90:075109
54. Brechtel E, Cordier G, Schäfer H (1981) J Less Common Metals 79(1):131
55. Benz R, Zachariasen W (1970) Acta Crystallogr Sect B Struct Crystallogr Cryst Chem 26(6):823
56. Charvillat J, Zachariasen W (1977) Inorg Nucl Chem Lett 13(3):161
57. Nuss J, Jansen M (2009) J Alloy Compd 480(1):57
58. Mizoguchi H, Hosono H (2011) J Am Chem Soc 133(8):2394
59. Nuss J, Jansen M (2012) Z Anorg Allg Chem 638(3–4):611
60. Shannon RT (1976) Acta Crystallogr Sect A Cryst Phys Diffr Theor Gen Crystallogr 32(5):751
61. Zintl E (1939) Angewandte Chemie 52(1):1
62. Papoian GA, Hoffmann R (2000) Angew Chem Int Ed 39(14):2408
63. Deller K, Eisenmann B (1976) ZEITSCHRIFT FUR NATURFORSCHUNG SECTION BA JOURNAL OF CHEMICAL SCIENCES 31(1):29
64. Kim H, Kang CJ, Kim K, Shim JH, Min BI (2015) Phys Rev B 91:165130
65. Magdysyuk OV, Nuss J, Jansen M (2013) Acta Crystallogr Sect B Struct Sci Cryst Eng Mater 69(6):547
66. Kim H, Kang CJ, Kim K, Shim J, Min B (2016) Phys Rev B 93(12):125116
67. Park J, Lee G, Wolff-Fabris F, Koh Y, Eom M, Kim Y, Farhan M, Jo Y, Kim C, Shim J et al (2011) Phys Rev Lett 107(12):126402
68. Feng Y, Wang Z, Chen C, Shi Y, Xie Z, Yi H, Liang A, He S, He J, Peng Y et al (2014) Sci Rep 4:5385
69. Wang K, Graf D, Lei H, Tozer S, Petrovic C et al (2011) Phys Rev B 84(22):220401
70. Abrikosov AA (1998) Phys Rev B 58:2788
71. He J, Wang D, Chen G (2012) Appl Phys Lett 100(11):112405
72. Wang K, Wang L, Petrovic C (2012) Appl Phys Lett 100(11):112111
73. Wang K, Graf D, Petrovic C (2013) Phys Rev B 87:235101
74. Masuda H, Sakai H, Tokunaga M, Yamasaki Y, Miyake A, Shiogai J, Nakamura S, Awaji S, Tsukazaki A, Nakao H et al (2016) Sci Adv 2(1):e1501117
75. Li L, Wang K, Graf D, Wang L, Wang A, Petrovic C (2016) Phys Rev B 93:115141
76. Guo Y, Princep A, Zhang X, Manuel P, Khalyavin D, Mazin I, Shi Y, Boothroyd A (2014) Phys Rev B 90(7):075120
77. Ray SJ, Alff L (2016) Phys Status Solidi (B)
78. Borisenko S, Evtushinsky D, Gibson Q, Yaresko A, Kim T, Ali M, Buechner B, Hoesch M, Cava RJ (2015) arXiv:1507.04847
79. Luo W, Xiang H (2015) Nano Lett 15(5):3230
80. Jung MH, Lacerda AH, Takabatake T (2002) Phys Rev B 65:132405
81. Han F, Wan X, Phelan D, Stoumpos CC, Sturza M, Malliakas CD, Li Q, Han TH, Zhao Q, Chung DY, Kanatzidis MG (2015) Phys Rev B 92:045112
82. Mizoguchi H, Matsuishi S, Hirano M, Tachibana M, Takayama-Muromachi E, Kawaji H, Hosono H (2011) Phys Rev Lett 106(5):057002
83. Kim SW, Lee K, Adroja D, Demmel F, Taylor J, Jung MH (2014) J Appl Phys 116(7):073901
84. Rosa P, Jesus C, Adriano C, Fisk Z, Pagliuso P (2015) J Phys Conf Ser. IOP Publishing: 012063
85. Fujimori Y, Kan SI, Shinozaki B, Kawaguti T (2000) J Phys Soc Jpn 69(9):3017
86. Zhu X, Lei H, Petrovic C, Zhang Y (2012) Phys Rev B 86:024527
87. Retzlaff R, Buckow A, Komissinskiy P, Ray S, Schmidt S, Mühlig H, Schmidl F, Seidel P, Kurian J, Alff L (2015) Phys Rev B 91:104519
88. Vinod K, Bharathi A, Satya A, Sharma S, Devidas T, Mani A, Sinha A, Deb S, Sridharan V, Sundar C (2014) Solid State Commun 192:60
89. Zołnierek Z, Troć R (1978) J Magn Magn Mater 8(3):210

Chapter 2
Experimental Techniques

2.1 Sample Preparation

2.1.1 Pulsed Laser Deposition

Pulsed laser deposition (PLD) is widely utilized to obtain a high-quality oxide epitaxial thin film [1]. Figure 2.1 shows a schematic image of the PLD method. In a vacuum chamber, the pulsed laser of KrF excimer ($\lambda = 248$ nm) is irradiated on the sintered target of the sample. The target material is vaporized and deposited on the substrate fixed at a face-to-face position. The substrate temperature is controllable by using infrared semiconductor laser up to $\sim$1000 °C. In addition, the background pressure is adjustable by introducing gases such as O_2 and Ar, although the deposition was conducted at high vacuum without any gas introduction in this study. The PLD chamber is equipped with the reflection high-energy electron diffraction (RHEED) system which enables in situ monitoring of the thin film growth: the single crystalline sample yields in the streak or dot pattern while the amorphous film projects a halo pattern according to their surface structures.

2.1.2 Sputtering

Constituent atoms of the solid are ejected from the target surface when a high-energy gas ion collides with the target. This phenomenon is called sputtering evaporation or just sputtering [2, 3]. Currently, sputtering is widely used in the semiconductor and spintronic device manufacturing process such as surface cleaning, micromachining, thin film and junction formation, and surface analysis. It is also applied to the thin film growth in basic research since it is possible to fabricate the large area thin film in addition to no limitation of material selection.

R. Sei, *Two-Dimensional Superconductivity in Rare Earth Oxybismuthides with Unusual Valent Bismuth Square Net*, Springer Theses,
https://doi.org/10.1007/978-981-99-7313-2_2

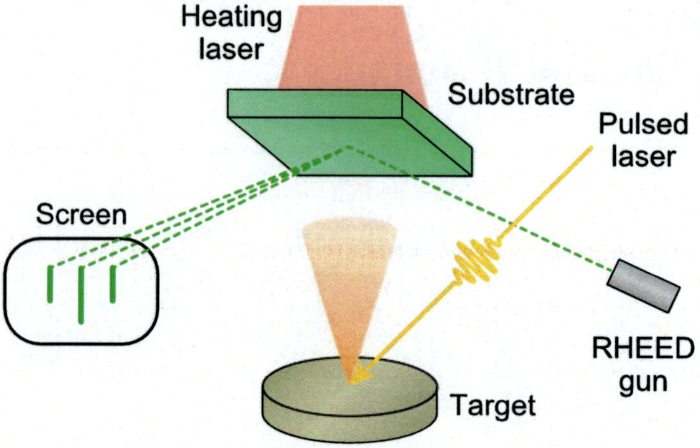

Fig. 2.1 A schematic image of PLD equipment

Sputtering generates a glow discharge. Its generation mechanism is as follows. Electrons accelerated to the anode by the electric field collide with the gas molecules such as Ar to generate the cations. The cations are accelerated to collide with the target on the cathode resulting in the emission of secondary electrons. The glow discharge is generated during a steady state by repeating the above process. When the cations collide with the target on cathode, not only secondary electrons but also the substances of the target on the cathode are emitted from the surface. These sputtered particles are deposited on the substrate and form the thin film.

Figure 2.2 shows a schematic image of the sputtering chamber. Sputtering is performed by introducing Ar gas into the vacuum chamber. The target material is evaporated by the irradiation of the ionized Ar and deposited on the substrate positioned in the opposite side. This chamber is equipped with a substrate heating laser, so thin film growth can be performed at $\sim$1000 °C maximum. There are two types of power sources: direct current (DC) and radio frequency (RF) modes. In DC sputtering, the target material is applied to the negative bias and the substrate is placed on the anode side. This approach is utilized in conducting materials such as a metal. However, DC sputtering cannot be applied to insulating materials due to the positively charged target by the collision with the cations. This charge-up can be suppressed by using RF power source because electrons and cations collide alternately by applying RF bias. The mobility of electrons in plasma is larger than that of cations, so the target surface is biased to a negative DC potential. RF sputtering can be used for the thin film growth of any materials from conductor to insulator.

In general, the deposition rate of the sputtering is slow due to the scattering of the sputtered particles in the high operating pressure. Magnetron sputtering is effective to increase the deposition rate. Magnets are arranged in order to produce a magnetic field line on the target surface as shown in Fig. 2.3. Secondary electrons are confined in the vicinity of the target along the magnetic field line, therefore it is possible to

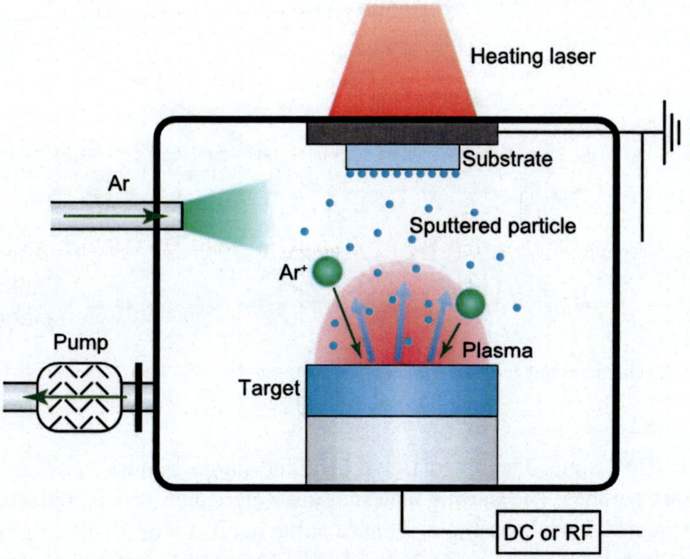

Fig. 2.2 A schematic image of sputtering chamber

Fig. 2.3 A schematic image of magnetron sputtering

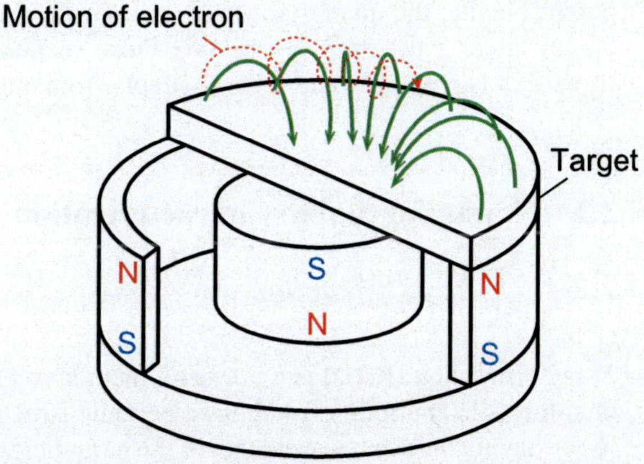

form a high-density plasma efficiently, resulting in the decrease of operating pressure. A disadvantage of the magnetron sputtering is the reduced efficiency of the use of target because of nonuniform sputtering at the target surface.

2.1.3 Vacuum Sealing

The polycrystalline powders of R_2O_2Bi cannot be obtained by the solid state reaction in air because of too excess amount of O in the atmosphere to achieve unusual

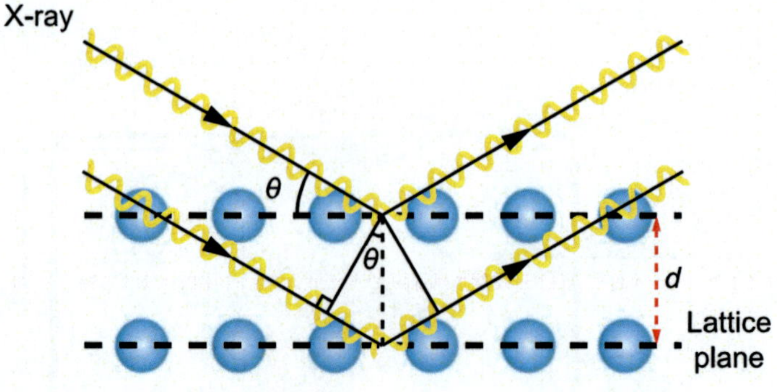

Fig. 2.4 A schematic image of Bragg's law

reductive Bi^{2-} state. Therefore, polycrystalline samples must be heated in vacuum sealed quartz tubes. The starting materials are weighed, mixed, pelletized, and sealed into the quartz tubes by using a vacuum valve in an Ar- or N_2-filled glovebox (O_2, $H_2O < 1$ ppm) in order to avoid the oxidation. The sample pellets are covered with Ta foil (0.04 mm thick; purity: 3Nup) to suppress the side reactions and contamination with Si. Then, the quartz tubes are transferred to air and connected to the rotary pump. Finally, the quartz tubes are sealed with a city/propane and O_2 mixed gas burner under evacuation. By heating those vacuum sealed samples in an electric furnace, it is possible to limit the O supply from only starting materials and quartz.

2.2 Crystallographic Characterization

2.2.1 X-Ray Diffraction

X-ray diffraction (XRD) is a powerful method to characterize the crystal structure of solids [4]. The solid crystal has a periodic structure with angstrom order. When X-ray having the same wavelength as the periodicity is incident on the crystal, X-ray is diffracted from the lattice plane according to the following Blagg's law (Fig. 2.4):

$$2d \sin\theta = n\lambda \tag{2.1}$$

where d is the distance between lattice planes, θ is the incident angle of X-ray, n is the integer, and λ is the wavelength of X-ray. Each substance possess its own diffraction pattern including the structural information such as constituent elements, lattice constants, and Miller indices. Therefore, XRD measurement enables the analysis of a detailed crystal structure and the identification of the substance.

Figure 2.5 shows a schematic image of the diffractometer utilized in this study (D8 DISCOVER, Bruker AXS). The equipment owns four movable axes: ω corresponds

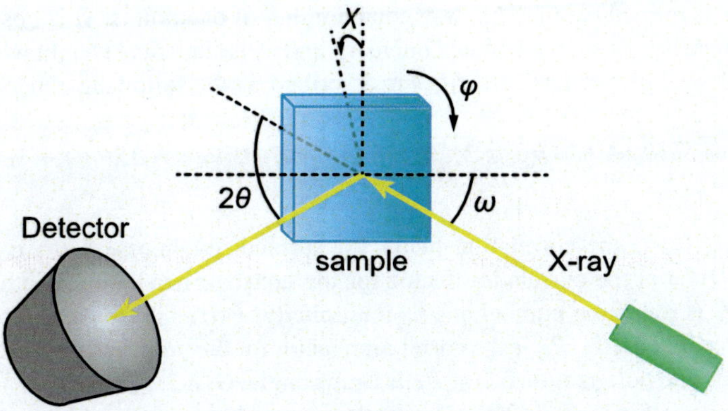

Fig. 2.5 A schematic image of measurement configuration of XRD equipment

to the incident angle of X-ray, 2θ corresponds to the angle between incident and diffracted X-ray, and ϕ and χ correspond to tilt angles along in-plane and out-of-plane directions to the sample surface, respectively. In general 2θ-θ scan, the diffraction pattern is corrected by moving 2θ with fixed $\omega = \theta$. As a result, all Bragg diffraction peaks can be corrected in polycrystalline samples, while only the diffraction peaks of a specific plane orientation (hkl) are detected in single crystalline samples by adjusting ϕ and χ axes. The rocking curve measurement is performed to evaluate the crystallinity of the crystal. The 2θ, ϕ, and χ axes are fixed in order to observe a specific Bragg hkl peak, then only ω is scanned. The smaller full width at half maximum (FWHM) of diffraction peak corresponds to the higher crystallinity. The ϕ scan is conducted to evaluate rotational symmetry and plane orientation of the crystal by fixing ω, 2θ, and χ. In this study, Cu Kα (λ =1.5418 Å) or monochromated Cu $K\alpha_1$ radiation (λ =1.5405 Å) was used for the measurement.

2.2.2 Rietveld Refinement

The Rietveld refinement is a powerful technique to analyze the structural parameters of solids such as lattice constants, fractional coordinates, and atomic displacement parameters [4, 5]. The measured powder X-ray or neutron diffraction pattern is refined by the theoretical pattern of the assumed structural model by the nonlinear least-squares method. In other words, the Rietveld refinement minimizes the residual sum of squares $S(\boldsymbol{x})$ represented by the following equation:

$$S(\boldsymbol{x}) = \sum_{i=1}^{N} w_i [y_i - f_i(\boldsymbol{x})]^2 \tag{2.2}$$

where $\boldsymbol{x}$ is refined parameters, N is total number of datapoints, y_i is observed measured intensity, $f_i(\boldsymbol{x})$ is calculated intensity, and w_i is statistical weight which usually adopts $1/y_i$. Furthermore, the $f_i(\boldsymbol{x})$ is described by the following equation:

$$f_i(\boldsymbol{x}) = S_{\mathrm{R}}(\theta_i)A(\theta_i)D(\theta_i)s\sum_K m_K|F(\boldsymbol{h}_K)|^2 P_K L(\theta_K)G(\Delta 2\theta_{iK}) + y_b(2\theta_i) \quad (2.3)$$

where $S_{\mathrm{R}}(\theta_i)$ is the correlation factor for surface roughness, $A(\theta_i)$ is absorption factor, $D(\theta_i)$ is the correlation factor for the constant irradiation width, s is scale factor, K is reflection number, m_K is multiplicity, $F(\boldsymbol{h}_K)$ is structure factor, $\boldsymbol{h}_K$ are reflection indices *hkl*, P_K is the correlation factor for the preferred orientation, $L(\theta_K)$ is Lorentz and polarization factor, θ_K is Bragg angle, $G(\Delta 2\theta_{iK}) = G(2\theta_i) - G(2\theta_K)$ is profile function, and $y_b(2\theta_i)$ is background correlation. Although the $f_i(\boldsymbol{x})$ seems to have a complicated formula, most of the factors are automatically and uniquely determined, and generally there are only four parameters which need to be refined: s, $y_b(2\theta_i)$, $F(\boldsymbol{h}_K)$, and $G(\Delta 2\theta_{iK})$. Hereafter, I explain each function used in the Rietan-FP program [6].

2.2.2.1 Scale Factor *s*

The scale factor s just multiplies the calculated intensity to fit into the observed intensity. The quantitative analysis in a multiphase system is based on the s values of each phase.

2.2.2.2 Background Correlation $y_b(2\theta_i)$

There is no method to estimate the background accurately, so the function of the Bragg angle is widely utilized to approximate the observed background. The Rietan-FP adopts the sum of Legendre polynomials as the fitting function for the background described by the following equation:

$$y_b(2\theta_i) = \sum_{j=0}^{11} b_j P_j(q_i) \quad (2.4)$$

$$P_j(q_i) = \left(\frac{2j-1}{j}\right) q_j P_{j-1}(q_j) - \left(\frac{j-1}{j}\right) P_{j-2}(q_i) \quad (2.5)$$

where $P_j(q_i)$ is the Legendre polynomial and b_j is background parameter. The $y_b(2\theta_i)$ is fitted by refining coefficients b_j although there is no physical meaning.

2.2.2.3 Crystal Structure Factor $F(\boldsymbol{h}_K)$

The $F(\boldsymbol{h}_K)$ is the main component to determine the peak intensity of each reflection and is described by the following equation:

$$F(\boldsymbol{h}_K) = \sum_j f_j g_j T_j \exp[2\pi i(hx_j + ky_j + lz_j)] \tag{2.6}$$

where j denotes the atom in unit cell, f_j is atomic form factor, g_j is occupancy, T_j is the Debye–Waller factor, and x_j, y_j, z_j are fractional coordinates. The f_j is the scattering power of each atom which is almost proportional to the atomic number, derived by the Fourier transform of the electron density:

$$f\left(\frac{\sin\theta}{\lambda}\right) \simeq f_0\left(\frac{\sin\theta}{\lambda}\right) = \int_{\text{atom}} \rho(\boldsymbol{r}) \exp[2\pi i(\boldsymbol{k} - \boldsymbol{k}_0)\cdot\boldsymbol{r}]\mathrm{d}v \tag{2.7}$$

where $\rho(\boldsymbol{r})$ is the electron density of the atom at $\boldsymbol{r}$, $\boldsymbol{k}$ and $\boldsymbol{k}_0$ are wave number vectors of scattered and incident waves, respectively. The f becomes small with increasing θ which is the main reason of 2θ dependence of diffraction intensity. The T_j indicates the attenuation of scattering amplitude due to the static disorder and thermal oscillation of the atoms, described by the following equation:

$$T_j = \exp\left[-B_j\left(\frac{\sin\theta_K}{\lambda}^2\right)\right] \tag{2.8}$$

where B_j is atomic displacement parameter indicating the mean square displacement from the mean position. The $F(\boldsymbol{h}_K)$ is calculated by refining structural parameters such as lattice constants, occupancy, fractional coordinates, and atomic displacement parameters, demonstrating the crystal structure of the sample.

2.2.2.4 Profile Function $G(\Delta 2\theta_{iK})$

The pseudo-Voigt function is widely used to fit the peak shape, which is a linear combination between the Lorentzian and Gaussian described by the following equation:

$$G(\Delta 2\theta_{iK}) = \eta\frac{2}{\pi H_K}\left[1 + 4\left(\frac{\Delta 2\theta_{iK}}{H_K}\right)^2\right]^{-1} + (1-\eta)\frac{2\sqrt{\ln 2}}{\sqrt{\pi}H_K}\exp\left[-4\ln 2\left(\frac{\Delta 2\theta_{iK}}{H_K}\right)\right] \tag{2.9}$$

where H_K is FWHM and η is the Lorentzian fraction. The $G(\Delta 2\theta_{iK})$ varies from Lorentzian to Gaussian by changing η. The crystalline size and microstrain affect both Lorentzian and Gaussian components of peak shapes, so these information can be obtained from the refined result of profile functions.

2.2.2.5 Reliability Indices

In order to judge the convergence of the Rietveld refinement, there are several indicators called reliability indices. The most important one is R_{wp} defined by the following equation:

$$R_{\mathrm{wp}} = \left\{ \frac{\sum_{i=1}^{N} [y_i - f_i(\boldsymbol{x})]}{\sum_{i=1}^{N} w_i y_i^2} \right\}^{\frac{1}{2}} \tag{2.10}$$

where the numerator corresponds to the residual sum of squares $S(\boldsymbol{x})$. Therefore, the Rietveld refinement is performed to minimize the R_{wp}. The R_{e} is the theoretical minimum of R_{wp} where the statistical deviation of the observed intensity is included, and it is described by the following equation:

$$R_{\mathrm{e}} = \left\{ \frac{N - P}{\sum_{i=1}^{N} w_i y_i^2} \right\}^{\frac{1}{2}} \tag{2.11}$$

where P is the number of refined parameters. Accordingly, the ratio of R_{wp} to R_{e}, i.e., $S = R_{\mathrm{wp}}/R_{\mathrm{e}}$ is utilized to judge the reliability of refined results: $S = 1$ indicates the perfect fitting, and $S < 1.3$ is desired to be regarded as a sufficiently reliable result. However, the larger observed intensity results in the larger S, so it is needed to cross-check the convergence of the refinement, e.g., by checking the reproducibility of the result from the different initial parameters.

2.2.3 *Atomic Force Microscope*

Atomic force microscopy (AFM) which is one of the scanning probe microscopy is used to evaluate the surface morphology of solids [7]. We can get the information about the surface structure in the order of nanometers by utilizing the atomic force between the solid surface and probe called cantilever. There is no restriction on the measurable sample for AFM because atomic force always exists between two objects in proximity.

Measurement principle of AFM is shown in Fig. 2.6. In order to detect the change of the cantilever position, the optical lever method is widely applied. Diode laser is irradiated on the back-miller of the cantilever, and angular variation of the reflected light is detected by a photodetector to evaluate the deflection of the cantilever. Measurement is performed by fixing the deflection and moving sample stage up and down in the feedback control. In this study, E-Sweep (Hitachi Nanotechnology Science) was used to observe the surface morphology of thin film samples.

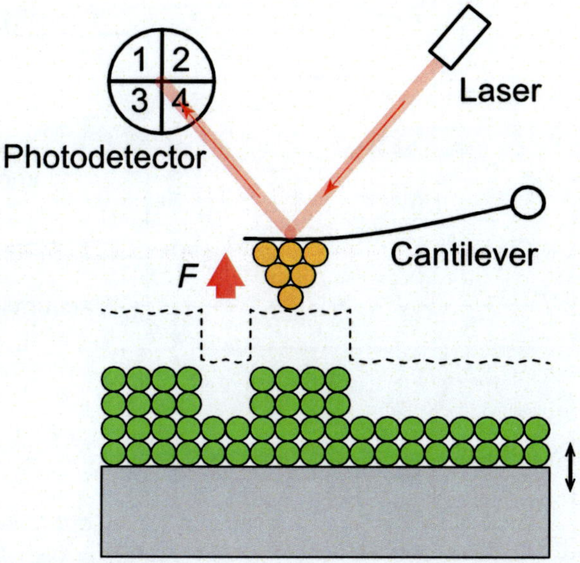

Fig. 2.6 A schematic image of AFM

2.3 Composition Analysis

2.3.1 X-Ray Photoemission Spectroscopy

When monochromatic radiation is irradiated to the substance, electrons are emitted by the photoelectric effect. These electrons are called photoelectrons. X-ray photoemission spectroscopy (XPS) is a technique which measures the energy spectrum of photoelectron by utilizing X-ray as a probe [8].

The photoelectric effect is represented by the following equation:

$$E_{\mathrm{K}} = h\nu - E_{\mathrm{B}} - \phi \tag{2.12}$$

where E_{K} is the kinetic energy of emitted photoelectron, $h\nu$ is the energy of incident photon, E_{B} is the binding energy of emitted photoelectron, and ϕ is the work function. A schematic image of a relationship between the photoelectric effect and XPS spectrum is shown in Fig. 2.7. Since a spectrometer and conductive samples possess the same Fermi level in the actual measurement, E_{B} is corrected to the following equation:

$$E_{\mathrm{B}} = h\nu - E_{\mathrm{K}} - \phi_{\mathrm{eq}} \tag{2.13}$$

where ϕ_{eq} is the work function of an equipment. Therefore, E_{B} can be derived from Eq. (2.13) by using a monochromatic light source which has a constant $h\nu$ and pre-measured ϕ_{eq}. In the case of the measurement of insulating samples, charge-

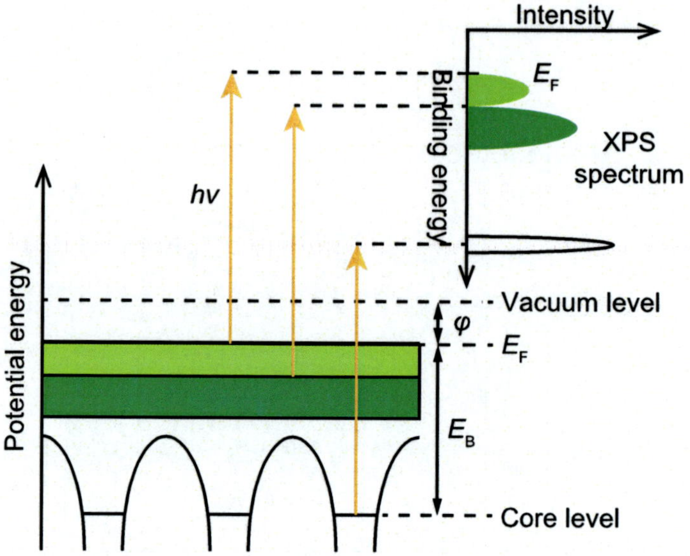

Fig. 2.7 A schematic image of photoelectric effect and XPS spectrum

up hampers the correct evaluation of the Fermi level. It is needed to prevent the sample from charge-up by a depositing conductive film and/or irradiating electron neutralizing beam.

As can be seen in Fig. 2.7, XPS spectrum has the information of the electrons in both core level and valence band. Each element has the characteristic E_B for each orbital electron, leading to the identification of the elements from peak positions of XPS spectrum. Quantitative analysis is also possible from peak intensity. E_B is shifted due to the surrounding chemical conditions, so the value of this chemical shift represents the valence state of the element. In this way, qualitative, quantitative, and chemical state analyses can be performed by XPS.

XPS is highly surface-sensitive because most photoelectrons lose their energy due to the scattering in the solid internal and are absorbed into the solid. We can obtain the information in the depth direction by sputtering the sample surface. In this study, XPS measurements was performed by using Al Kα radiation as the probe in PHI 5000 VersaProbe (Ulvac-Phi, Inc.) equipped with sputtering system.

2.3.2 *Electron Probe Microanalyzer*

Electron probe microanalyzer is a powerful tool to perform qualitative and quantitative analyses of constituent elements in the solid [9]. When accelerated electrons with energy 1–20 keV are irradiated on the surface of the substance, inner shell electrons are excited to generate holes in the orbit. Then, outer shell electrons fall into this empty orbit, resulting in the generation of characteristic X-ray. Since the energy of the characteristic X-ray has a unique value for each element, qualitative analysis can be performed by detecting it. In addition, quantitative analysis is performed from the intensity of characteristic X-ray. Mapping analysis is also possible by scanning the electron probe two-dimensionally.

Energy dispersive X-ray spectrometer (EDX) is widely utilized to detect the characteristic X-ray as follows. The incident X-ray into the semiconductor device generates electron–hole pairs whose numbers are proportional to the energy of the X-ray. These electrons are amplified by a strong electric field, resulting in pulses with intensity proportional to the energy of incident X-ray. In this study, the qualitative analysis of polycrystalline samples was mainly conducted by EDX installed in scanning electron microscopy (JEOL Ltd., JSM-7100F). The energy of electron probe was 15 keV.

2.3.3 *Inductively Coupled Plasma Mass Spectrometry*

Inductively coupled plasma (ICP) is a form of discharge caused by the induced current of the high-frequency coil [10]. When Ar gas flows through the discharging high-frequency coil, a large amount of ions and electrons, so-called plasma, are generated by the following chain reaction:

$$\begin{aligned} e^- + Ar &\rightarrow 2e^- + Ar^+ \\ 2e^- + 2Ar &\rightarrow 4e^- + 2Ar^+ \end{aligned} \tag{2.14}$$

where e^- denotes electrons. The temperature of the generated plasma reaches 10000 K, and it is used for atomization and thermal excitation of the sample. ICP mass spectrometry (ICP-MS) is a composition analysis method by utilizing this generated chemical species with mass spectrometer. The detection limit is very high as about ppb, although it is necessary to dissolve samples in acid for the analysis. In this study, ICP-MS was performed by Foundation for Promotion of Material Science and Technology of Japan.

2.4 Magnetic Measurements

Electromagnetic induction method is utilized to evaluate the magnetic moment of the material [11]. When the sample moves in a homogeneous magnetic field, the electromotive force V is induced in the pick-up coil due to the change of magnetic flux density Φ, that is, $V = -\mathrm{d}\Phi/\mathrm{d}t$. According to the relationship $\Phi = km$ where m is the magnetic moment of the sample and k is the factor of proportionality which

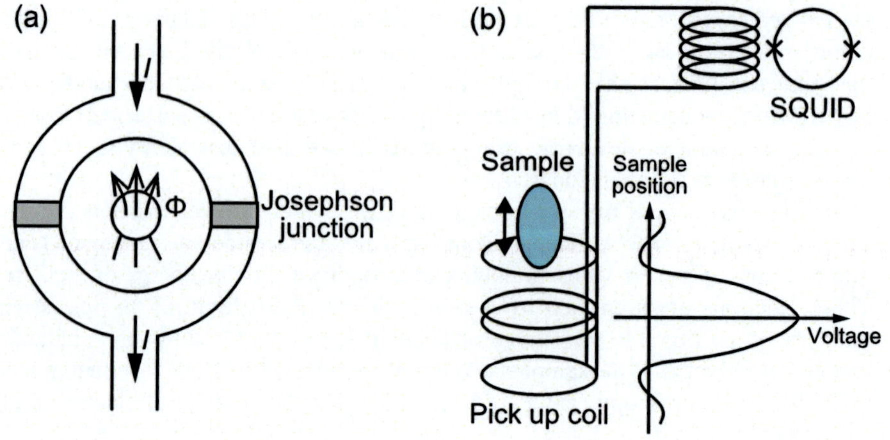

Fig. 2.8 **a** A schematic image of SQUID magnetometer. **b** Measurement configuration of MPMS. Inset shows schematic image of output voltage

depends on the measurement system, the m is obtained by the integral calculus of V:

$$m = \frac{1}{k}\int_0^t V\mathrm{d}t. \tag{2.15}$$

Superconducting quantum interference device (SQUID) is a powerful tool to evaluate the magnetic moment with high sensitivity [11, 12]. The SQUID magnetometer is composed of a superconducting ring where two Josephson junctions are connected in parallel (Fig. 2.8). In this system, the maximum of superconducting current is given by the following formula:

$$I_{\max} = 2I_c\left|\cos\pi\frac{\Phi}{\Phi_0}\right| \tag{2.16}$$

where I_c is the critical current of each Josephson junction and $\Phi_0 \approx 2.07 \times 10^{-15}$ Wb is the flux quantum. Since the current is a periodic function with the periodicity of Φ_0, high sensitive detection of the magnetic moment is possible with the magnetic quantum as a unit. In the actual measurement, a steady bias current larger than I_c flows in the SQUID ring, and thereby the voltage fluctuation with the periodicity of Φ_0 is detected.

In this study, the magnetic property measurement system (MPMS, Quantum Design, Inc.) equipped with a SQUID magnetometer was utilized to evaluate the magnetic properties. Figure 2.8b shows a schematic image of measurement configuration. The sample moved in the pick up coil and the changed magnetic flux is indirectly detected by the SQUID magnetometer. This is because the SQUID magnetometer needs to be shielded with the superconductor in order to avoid the influence from noise such as an external magnetic field.

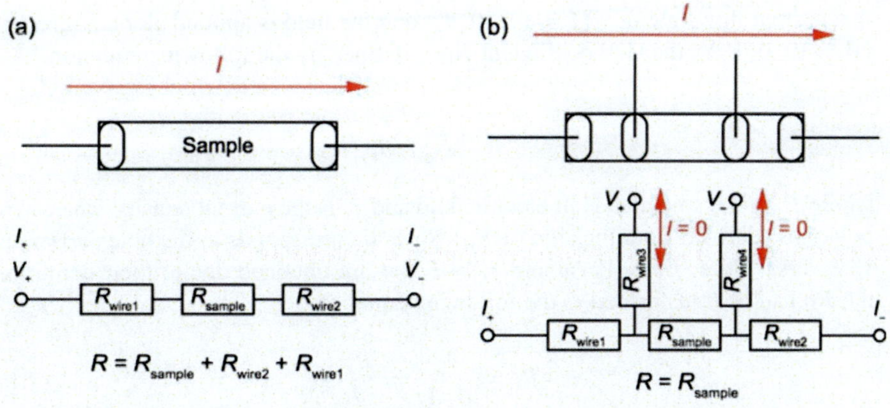

Fig. 2.9 Schematic images of **a** two-probe and **b** four-probe configuration. I: current; V: voltage

2.5 Electrical Measurements

2.5.1 Four-Probe Method

Electrical resistance $R = V/I$ can be evaluated by wiring current and voltage electrodes properly. However, if both current and voltage are measured in the same electrode, i.e., two-probe method, the obtained R contains contribution not only from the sample but also from wires and electrodes (Fig. 2.9a). Therefore, the four-probe method is better to evaluate correct R value of the sample where electrodes for current and voltage are separated (Fig. 2.9b) [13]. In this configuration, obtained R corresponds to the resistance between electrodes for voltage because current does not flow through the voltmeter in an ideal condition.

The obtained sample resistance R can be converted to electrical resistivity ρ by the following equation:

$$\rho = R\frac{wt}{L} \tag{2.17}$$

where w is a width of the sample perpendicular to the current direction, t is the sample thickness, and L is the distance between electrodes for voltage.

2.5.2 Hall Measurement

Hall measurement is a useful technique to derive carrier type, density, and mobility. When a magnetic field $\mu_0 \boldsymbol{H} = \boldsymbol{B}$ is applied perpendicular to the current $\boldsymbol{I}$, the charge carriers are accumulated in $\boldsymbol{I} \times \boldsymbol{B}$ direction due to the Lorentz force, resulting in the generation of electric field $\boldsymbol{E}$. This phenomenon is called the Hall effect. By assuming that current flows along x ($\boldsymbol{I} = (I_x, 0, 0)$) and the field is applied along z direction ($\boldsymbol{B} = (0, 0, B_z)$), the Hall coefficient R_{H} is defined by the following equation:

$$R_{\mathrm{H}} \equiv \frac{E_y}{j_x B_z} \tag{2.18}$$

where E_y is the so-called Hall electric field and j_x is the current density. The carrier polarity is derived from the sign of R_{H}: $R_{\mathrm{H}} > 0$ corresponds to the hole carrier and vice versa. Here, $E_y = V_y/w$ and $j_x = I_x/wt$ are obtained due to their definition; the R_{H} can be transformed to the following equation:

$$R_{\mathrm{H}} = \frac{V_y t}{I_x B_z} = \frac{R_{yx} t}{B_z} \tag{2.19}$$

where $R_{yx} = V_y/I_x$ is the Hall resistance. In addition, the carrier density n and mobility μ are represented by using R_{H} as follows:

$$n = \left| \frac{1}{R_{\mathrm{H}} e} \right| \tag{2.20}$$

$$\mu = \left| \frac{R_{\mathrm{H}}}{\rho} \right| \tag{2.21}$$

where e is the elementary charge. Therefore, we can obtain the carrier type, density, and mobility by measuring Hall resistance.

2.5.3 Measurement Configuration

Figure 2.10 shows a schematic image of measurement configuration in this study. Both film and polycrystalline samples were processed into a bar shape, and the electrodes were fabricated by Au deposition with an In crimp, only In crimp, or Ag paste. All electrical measurements were conducted by the physical property measurement system (PPMS, Quantum Design, Inc.) equipped with a sample rotator system. Temperature range was from 1.8 K without the rotator system or 1.85 K with it to 300 K, and the magnetic field was varied in the range of $-9 \leq \mu_0 H \leq 9$ T by a superconducting magnet. The direction of the applied magnetic field could be changed by rotating the samples in order to evaluate the angular dependence of MR.

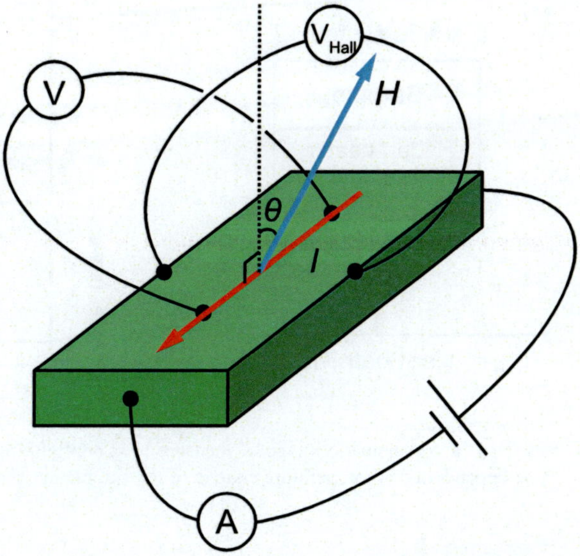

Fig. 2.10 A schematic image of electrical measurement configuration in this study

2.6 Specific Heat Measurements

Specific heat measurements were conducted by the thermal relaxation method [14] with a ^{3}He/^{4}He dilution refrigerator [15] with the help of the Kawaji group in Tokyo Institute of Technology.

2.6.1 Thermal Relaxation Method

Figure 2.11a shows the measurement configuration of the thermal relaxation method composed of the sample, holder, and heat bath. The holder equipped with heater and thermometer is connected to the heat bath by thermal resistance possessing thermal conductivity k. When the quantity of heat per unit time $Q = RI^2$ is applied to the sample system, the energy balance is represented by the following equation:

$$C_p \frac{\mathrm{d}T}{\mathrm{d}t} = Q - k(T - T_0) \tag{2.22}$$

where C_p is the heat capacitance of the sample system, T and T_0 are the temperature of sample system and heat bath, respectively. In a steady state denoted by process 2 in Fig. 2.11b, Eq. (2.22) is transformed into the relationship $Q = k(T - T_0) = k\Delta T$ due to the situation of $\mathrm{d}T/\mathrm{d}t = 0$, enabling to evaluate k. When the heat supply is stopped ($Q = 0$), temperature of the sample system gradually relaxes down to T_0 by following the relationship $C_p\mathrm{d}T/\mathrm{d}t = -k(T - T_0)$. By solving this differential

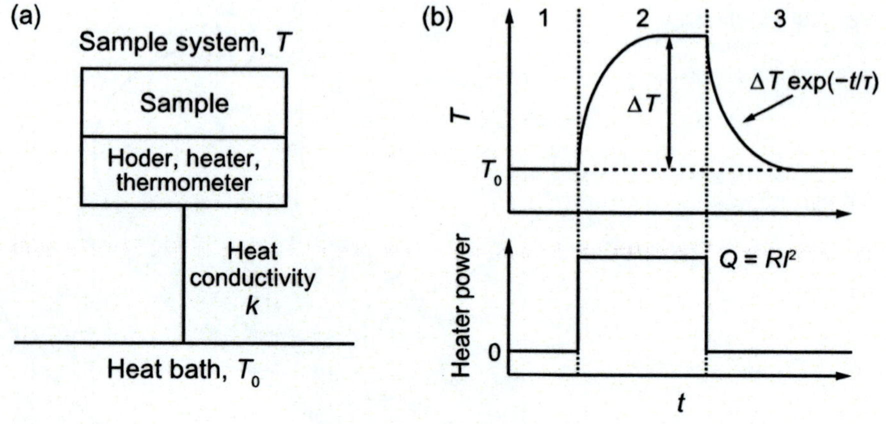

Fig. 2.11 **a** A schematic image of measurement configuration for thermal relaxation method. **b** Time dependence of temperature change of the sample system (top) and heater power (bottom)

equation, the function of the relaxation curve (process 3 in Fig. 2.11b) is obtained as follows:

$$T = \Delta T \exp\left(-k\frac{t}{C_p}\right) = \Delta T \exp\left(-\frac{t}{\tau}\right) \tag{2.23}$$

where $\tau = C_p/k$ is the time constant. Therefore, C_p can be derived from τ or the fitting of the relaxation curve. It is noted that the obtained C_p contains the contribution from both sample itself and holder, so it is necessary to subtract the latter signal measured beforehand.

2.6.2 Dilution Refrigerator

The ^{3}He/^{4}He dilution refrigerator is utilized in order to realize the extremely low temperature of 0.5 K or less. The He has two stable isotopes, i.e., ^{3}He which is Fermion and ^{4}He which is Boson. At a very low temperature, their mixed solution undergoes phase separation into two phases; c phase which contains rich ^{3}He (~100mol%) and d phase where ^{3}He is dilute (~6.4mol%). In a dilution refrigerator, these two phases are in contact. Here, ^{3}He has larger vapor pressure than ^{4}He, so it can be selectively vaporized in the d phase. This results in the melting of ^{3}He from the c phase into d phase with the endothermic process, enabling the continuous realization of extremely low temperature.

References

1. Ohashi S, Lippmaa M, Nakagawa N, Nagasawa H, Koinuma H, Kawasaki M (1999) Rev Sci Instrum 70(1):178
2. Wasa K, Hayakawa S (1988) Sputtering technology. Kyoritsu Shuppan
3. Kinbara A (1984) Sputtering phenomenon. University of Tokyo Press
4. Nakai I, Izumi F (eds) (2009) Practice of powder x-ray diffraction for analysis. Asakura Publishing
5. Tsubota M, Itoh T (2012) Rietveld refinement learned with RIETAN-FP. Johokiko
6. Izumi F, Momma K (2007) Applied crystallography XX, Solid state phenomena, vol 130, pp 15–20. Trans Tech Publications. https://doi.org/10.4028/www.scientific.net/SSP.130.15
7. TSSS of Japan (ed) (2002) Scanning probe microscope for nanotechnology. Maruzen publishing
8. TSSS of Japan (ed) (1998) X-ray photoelectron spectroscopy. Maruzen Publishing
9. Hobo T (ed) (1996) High purification technology system. Fuji tekunoshisutemu
10. TJS for Analytical Chemistry (ed) (2011) Handbook of analytical chemistry. Maruzen Publishing
11. TCS of Japan (ed) Experimental chemistry course. Maruzen Publishing
12. Grosso G, Parravicini GP (2005) Solid state physics. Yoshiokashoten
13. Ohtsuka Y, Kobayashi Y (eds) Experimental physics course. Maruzen Publishing
14. TCS of Japan (ed) Experimental chemistry course. Maruzen Publishing
15. PDE Board (ed) (2005) Physics dictionary. Baifukan

Chapter 3
Development of Solid-Phase Epitaxy Techniques

3.1 Introduction

Epitaxial thin film growth of $R_2O_2Bi$is a challenging issue because of the high vapor pressure and strongly reductive state of Bi^{2-}. Indeed, the PLD method was not applicable to the fabrication of Y_2O_2Bi epitaxial thin films on various lattice-matched substrates even in high vacuum ($< 4 \times 10^{-6}$), resulting in the preferential formation of cubic C-rare earth type $(Y,Bi^{3+})_2O_3$ epitaxial thin film (Fig. 3.1). In order to obtain Y_2O_2Bi epitaxial thin film, solid-phase epitaxy (SPE) is one promising technique because various layered oxide epitaxial thin films were successfully achieved for the first time via SPE [1, 2]. The main difficulties in the fabrication of Y_2O_2Bi epitaxial thin film are high vapor pressure and unusual reductive state of Bi. Thus, an ingenious approach overcoming the above problems will be desired.

In this chapter, two SPE methods were newly developed for Y_2O_2Bi: reductive SPE and multilayer SPE (Fig. 3.2). In the first method, Y_2O_2Bi epitaxial thin film was obtained through the direct reaction between Y_2O_3 amorphous thin film and (Y, Bi) mixed powders via two-step heating in a highly reductive condition, connecting to the suppression of oxidation and evaporation of Bi (Sect. 3.2). However, the low crystalline quality of the film hampered the evaluation of fundamental physical properties, especially the electrical conduction. Therefore, the SPE method was improved by utilizing a multilayered precursor (Sect. 3.3). The highly crystalline Y_2O_2Bi epitaxial thin film enabled to observe intrinsic physical properties, demonstrating the two-dimensional electronic state with strong spin–orbit coupling.

R. Sei, *Two-Dimensional Superconductivity in Rare Earth Oxybismuthides with Unusual Valent Bismuth Square Net*, Springer Theses,
https://doi.org/10.1007/978-981-99-7313-2_3

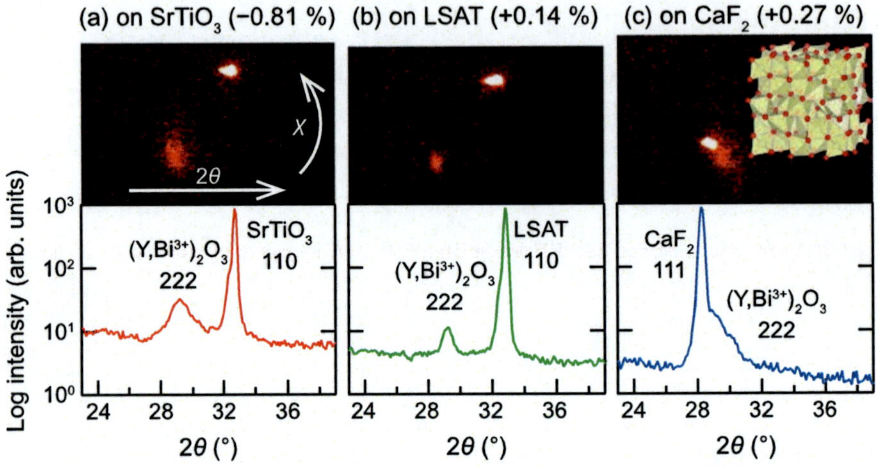

Fig. 3.1 Two-dimensional XRD patterns (upper) and their projections (bottom) of films fabricated by PLD method deposited on **a** $SrTiO_3$, **b** $(LaAlO_3)_{0.3}(SrAl_{0.5}Ta_{0.5}O_3)_{0.7}$ (LSAT), and **c** CaF_2 substrates. Numerals between brackets denote lattice mismatch. Inset of (**c**) shows crystal structure of cubic Y_2O_3

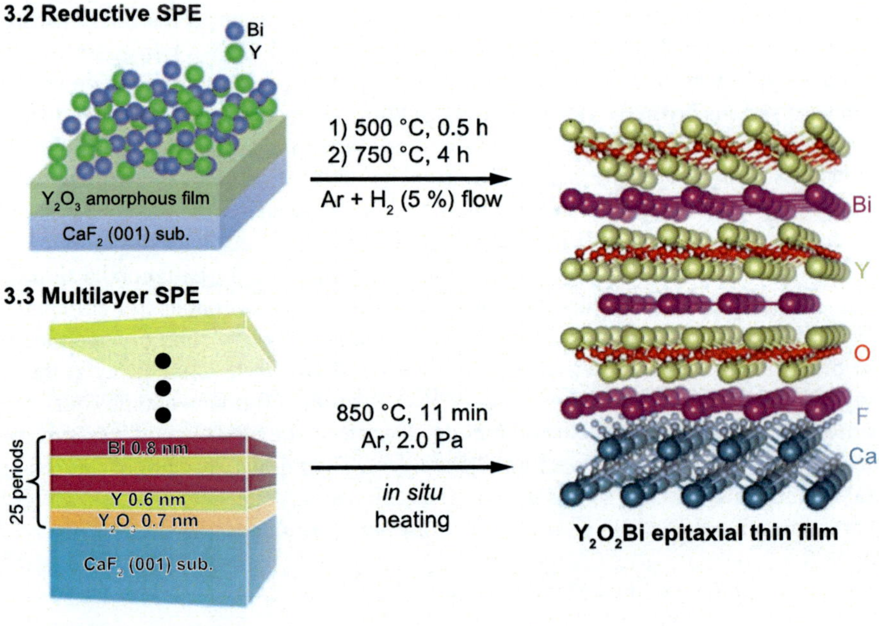

Fig. 3.2 Schematic images of newly developed two SPE methods

3.2 Development of Reductive Solid-Phase Epitaxy

3.2.1 Experimental

A schematic procedure[1] of reductive SPE is shown in Fig. 3.3. Y_2O_3 precursor films were deposited by the PLD method on CaF_2 (001) substrate ($a/\sqrt{2} = 3.863$ Å, 0.27% of lattice mismatch) at room temperature under high vacuum ($< 10^{-6}$ Pa). The typical thickness of Y_2O_3 film was ~ 40 nm without any XRD peaks, indicating amorphous nature. Then, Y and Bi (purity: 3N) mixed powders with a molar ratio of 1 : 1 were sandwiched between two Y_2O_3 amorphous thin films. The sandwiched structure was covered with Ta foil in order to fix the structure. This operation was conducted in an Ar-filled glovebox. Finally, the product was heated in two steps at 500 °C for 0.5 h and at 750 °C for 4 h in Ar + H_2 (5%) gas flow. Crystal structures of the fabricated films were characterized by XRD with a two-dimensional detector using Cu Kα radiation.

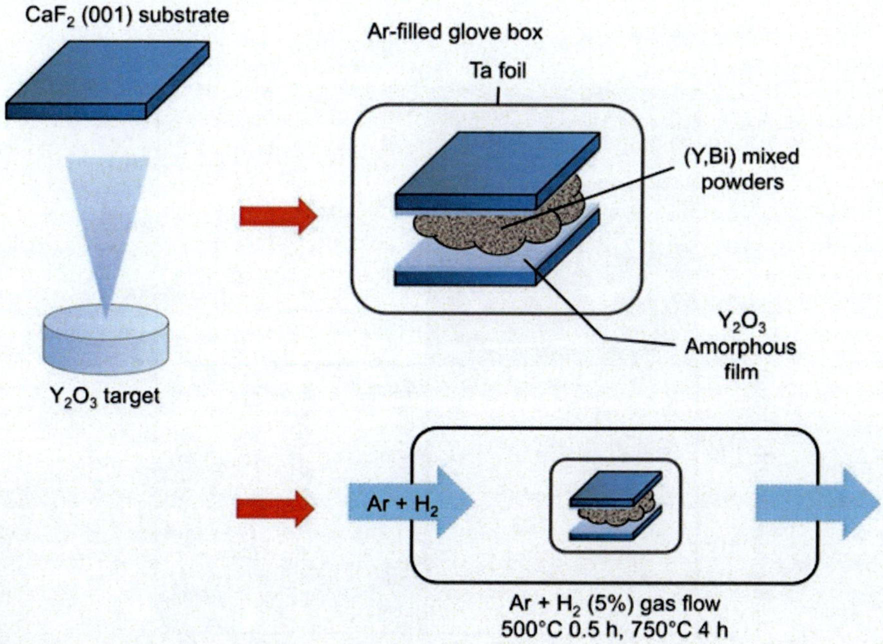

Fig. 3.3 A schematic procedure of reductive SPE

[1] This section contains the contents of the following publication. Reprinted with permission from R. Sei, T. Fukumura, T. Hasegawa, *Crystal Growth & Design* **14**, 4227 (2014). Copyright 2014 by American Chemical Society.

3.2.2 Fabrication Results

Figure 3.4a–c shows a typical two-dimensional XRD pattern and its projections of an asymmetric plane for fabricated film. A spot-like peak corresponding to Y_2O_2Bi 103 diffraction was clearly observed adjacent to CaF_2 111 diffraction peak, indicating the single crystalline growth of the Y_2O_2Bi phase. Furthermore, Y_2O_2Bi 00*l* diffraction peaks were observed along CaF_2 002 diffraction peak in the 2θ-θ XRD pattern of a symmetric plane (Fig. 3.4d), exhibiting *c*-axis oriented growth of the film. The calculated lattice constants were $a = 3.87$ Å and $c = 13.1$ Å, where the *c*-axis length showed a small contraction in contrast to that in the previous study ($a = 3.8734$ Å and $c = 13.2469$ Å) [3]. The reason will be discussed in Sect. 3.3. The epitaxial relationship between the film and the substrate was determined from Fig. 3.4a, d: Y_2O_2Bi [100] ∥ CaF_2 [110] and Y_2O_2Bi [001] ∥ CaF_2 [001]. These results represent the successful fabrication of Y_2O_2Bi epitaxial thin film with *c*-axis orientation. This is the first fabrication report of single crystalline R_2O_2Bi.

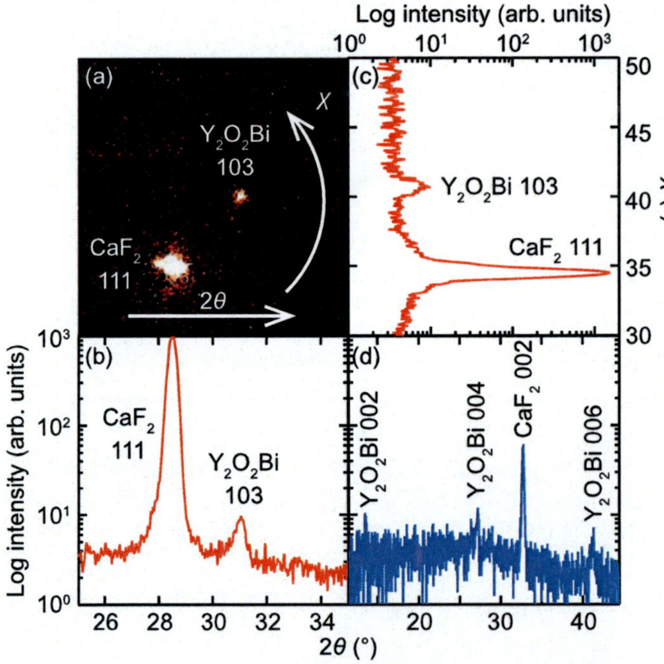

Fig. 3.4 XRD patterns of the film fabricated by reductive SPE. **a** Two-dimensional XRD pattern of the asymmetric plane and its projection along **b** 2θ and **c** χ axes. **d** XRD pattern of the symmetric plane projected along 2θ axis

3.2.3 Reaction Mechanism

In this study, a two-step heating process was utilized to grow Y_2O_2Bi epitaxial thin film. This technique was generally applied to the fabrication of polycrystalline R_2O_2Bi powders: mixed starting materials (Bi, R, and R_2O_3) with a stoichiometric composition were heated in an evacuated silica tube at 500 °C, followed by heating at 750 °C [3]. A role of the first heating step can be considered as the formation of rare earth bismuthides RBi referring to the R-Bi binary phase diagram [4]. Besides, other researchers have synthesized polycrystalline R_2O_2Bi and analogous R_2O_2Sb from the mixtures of Bi (Sb), RBi (RSb), and R_2O_3 [5–7]. These facts suggest that the formation of RBi was a key to achieve the R_2O_2Bi phase. Furthermore, Y_2O_3 film was kept as the amorphous state even after the first heating step, confirmed by solely heating the Y_2O_3 amorphous thin film under vacuum (Fig. 3.5); no RHEED image with the slight diffraction peak intensity indicates the mostly amorphous nature of Y_2O_3 film at the first heating step, followed by the complete crystallization at the second heating step. Therefore, the reaction process of the reductive SPE can be represented by the following formula:

$$\begin{aligned} &[\mathrm{Y_2O_3}\text{ amorphous thin film}] + [(\mathrm{Y,Bi})\text{ powders}] \\ &\xrightarrow[\mathrm{Ar+H_2}\ (5\,\%)\text{ flow}]{500\,^\circ\mathrm{C},\ 0.5\,\mathrm{h}} [\mathrm{Y_2O_3}\text{ amorphous thin film}] + \mathrm{YBi} \\ &\xrightarrow[\mathrm{Ar+H_2}\ (5\,\%)\text{ flow}]{750\,^\circ\mathrm{C},\ 4\,\mathrm{h}} [\mathrm{Y_2O_2Bi}\text{ epitaxial thin film}] \end{aligned} \tag{3.1}$$

In order to consider the details of the formation mechanism of Y_2O_2Bi epitaxial thin film, two referential experiments were conducted. The first topic was the replacement of precursor Y_2O_3 thin film from the amorphous one to epitaxial one. Other conditions such as the mixing ratio of (Y, Bi) powders and the heating process were unchanged. This reaction can be summarized by the following formula:

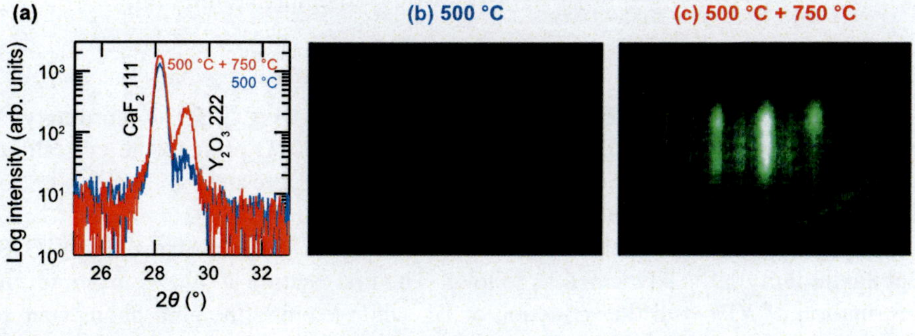

Fig. 3.5 **a** XRD pattern of asymmetric plane for Y_2O_3 film on CaF_2 substrate heated at 500 °C and then at 750 °C. RHEED patterns of Y_2O_3 film after heating at **b** 500 °C and then at **c** 750 °C, representing the mostly amorphous phase and the crystallized phase, respectively

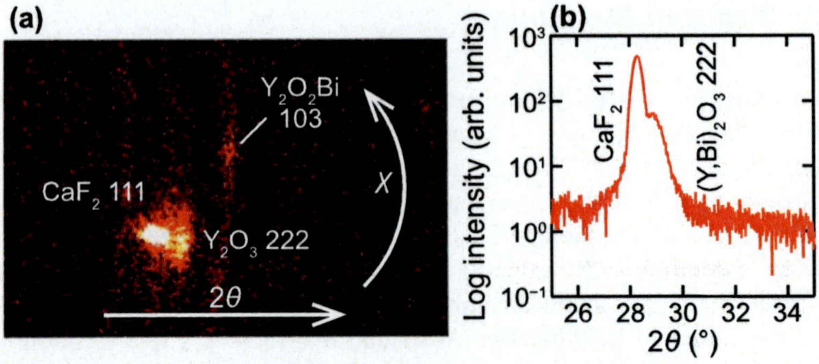

Fig. 3.6 **a** Two-dimensional XRD pattern of the film after heating Bi and Y powders sandwiched between Y_2O_3 (001) epitaxial films. **b** The XRD pattern along the 2θ direction of the films fabricated from only Bi powder and Y_2O_3 amorphous film on CaF_2 substrate

$$[Y_2O_3\text{ epitaxial thin film}] + [(Y,Bi)\text{ powders}] \xrightarrow[\text{Ar + H}_2\text{ (5 \%) flow}]{\text{1) 500 °C, 0.5 h, 2) 750 °C, 4 h}} [Y_2O_3\text{ epitaxial thin film}] + [Y_2O_2Bi\text{ polycrystalline thin film}] \quad (3.2)$$

Obtained film showed strong diffraction peak of single crystalline Y_2O_3 with a small amount of polycrystalline Y_2O_2Bi (Fig. 3.6a), implying that the growth of Y_2O_2Bi phase was hampered once Y_2O_3 was crystallized because of the high thermodynamical stability of Y_2O_3 phase.

The second topic was the role of Y. The film was fabricated from Y_2O_3 amorphous thin film and only Bi powder with the same heating condition, as represented by the following formula:

$$[Y_2O_3\text{ amorphous thin film}] + [\text{Bi powder}] \xrightarrow[\text{Ar + H}_2\text{ (5 \%) flow}]{\text{1) 500 °C, 0.5 h, 2) 750 °C, 4 h}} [(Y,Bi^{3+})_2O_3\text{ epitaxial thin film}] \quad (3.3)$$

Obtained film was $(Y,Bi^{3+})_2O_3$ epitaxial thin film (Fig. 3.6b). Accordingly, Y served as a reductant for Bi in the first heating at 500 °C prior to the crystallization of the Y_2O_3 amorphous film. In other words, the presence of Bi^{2-} might be indispensable for the subsequent formation of Y_2O_2Bi.

With the above results taken into account, the formation mechanism of the Y_2O_2Bi epitaxial thin film is advocated as follows. The first heating at 500 °C promotes the formation of YBi with the reduction of Bi, while keeping the amorphous state of the Y_2O_3 film. The second heating at 750 °C stimulates the epitaxial growth of Y_2O_2Bi with the help of the lattice-matched CaF_2 substrate. Hence, the first heating at 500 °C is indispensable to circumvent the formation of Y_2O_3 prior to the formation of Y_2O_2Bi.

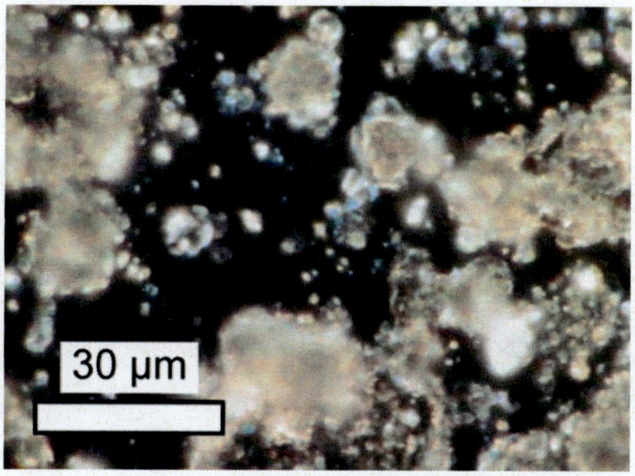

Fig. 3.7 Dark field optical microscope image of the film fabricated by reductive SPE

3.2.4 Drawbacks of Reductive Solid-Phase Epitaxy

It was difficult to measure the physical properties and to evaluate them properly due to some problems although Y_2O_2Bi epitaxial thin film was successfully obtained for the first time by developing reductive SPE. Figure 3.7 shows a dark field optical microscope image of the film fabricated by the reductive SPE method, indicating the large amount of powder residue at the film surface. In addition, weak intensity of XRD peak (Fig. 3.4) indicated quite a small crystalline volume of Y_2O_2Bi phase. These drawbacks hampered the fundamental evaluation of physical properties. A novel fabrication technique is desired to improve the crystalline quality and to eliminate surface residue.

3.3 Development of Multilayer Solid-Phase Epitaxy

3.3.1 Experimental

In order to overcome[2] the above problems of reductive SPE, all of the reaction precursor was converted to the thin film form (Fig. 3.8). The multilayered precursor composed of Bi (0.8 nm thick), Y (0.6 nm), and Y_2O_3 (0.8 nm) layers was deposited on cubic CaF_2 (001) single-crystal substrate (lattice mismatch: 0.27%) at ambient temperature by magnetron sputtering in Ar gas atmosphere as shown in Fig. 3.8. The subnanometer thick layers were crucial to enhance reaction efficiency and to preserve

[2] This section contains the contents of the following publication. Reprinted with permission from R. Sei, T. Fukumura, T. Hasegawa, *ACS Applied Materials & Interfaces* **7**, 24998 (2015). Copyright 2015 by American Chemical Society.

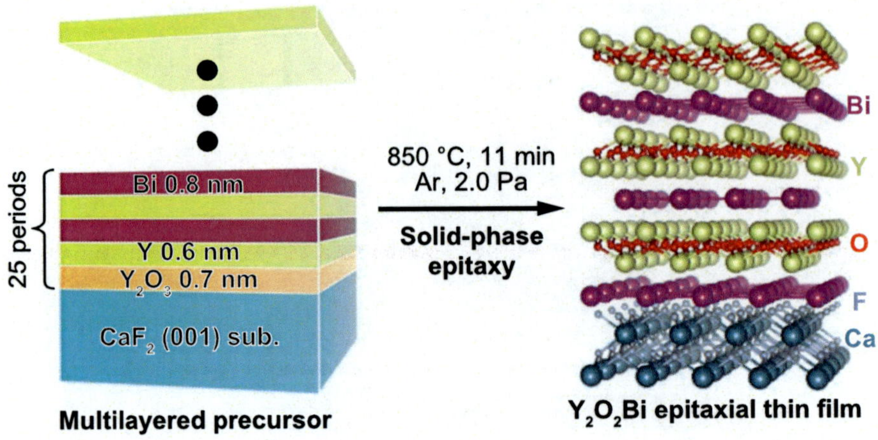

Fig. 3.8 A schematic image of multilayer SPE. Right panel shows the crystal structure of Y_2O_2Bi (001) epitaxial thin film on CaF_2 (001) substrate

flat film morphology. The base and working pressure of the sputtering chamber were less than 3.3×10^{-5} and 2.0 Pa, respectively. Each layer was deposited from pure Bi (4Nup), Y (3N), and Y_2O_3 (3N) targets in a sequence of [Y_2O_3/Y/Bi/Y/Bi] as a multilayer unit with 25 periods, followed by the deposition of Y cap layer because of volatile Bi. Subsequently, the precursor was in situ heated with a heating rate of 30 °C up to 850 °C, and then the temperature was held for 11 min. The typical film thickness was around 90 nm.

Crystal structure was evaluated by XRD using Cu Kα_1 radiation. Surface morphology was observed by AFM. Electronic state was evaluated by XPS using Al Kα radiation in ultrahigh vacuum (1×10^{-6} Pa) after surface cleaning by in situ Ar sputtering. Samples were mounted on the sample holder by using a conducting carbon tape in order to prevent the sample from charge-up. In addition, the electron beam was irradiated for charge neutralization during the measurement. The XPS peak positions were calibrated by C 1 s peak position (284.8 eV) [8]. Electronic transport properties were measured by standard four-probe method by PPMS equipped with a sample rotator system.

3.3.2 Crystal Structure and Valence State of Bi

Figure 3.9a shows XRD out-of-plane 2θ-θ scan of the fabricated thin film. Sharp Y_2O_2Bi 00l peaks without any secondary phases were observed in contrast to the result in reductive SPE. The lattice constants were a = 3.880 Å and c = 13.129 Å. The a-axis length was similar to that of polycrystalline powder (a = 3.8734 Å), whereas the c-axis length was slightly shorter (c = 13.2469 Å) [3] as discussed

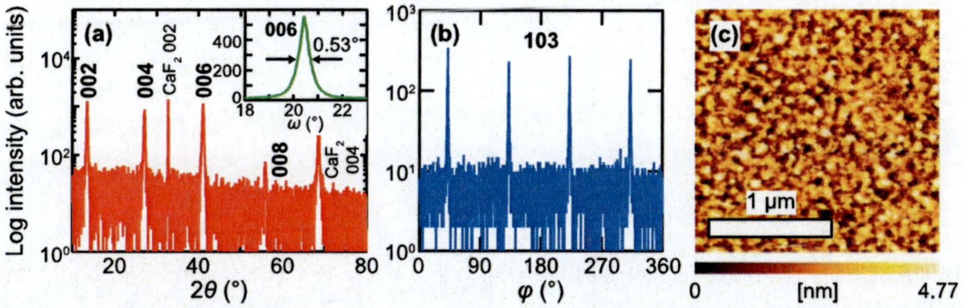

Fig. 3.9 a XRD out-of-plane 2θ-θ scan of the Y_2O_2Bi epitaxial thin film on the CaF_2 substrate. Inset shows 006 rocking curve. b XRD in-plane ϕ scan of 103 peak. $\phi = 0°$ corresponded to the CaF_2 (100) plane. c Atomic force microscope planar image of the Y_2O_2Bi epitaxial thin film

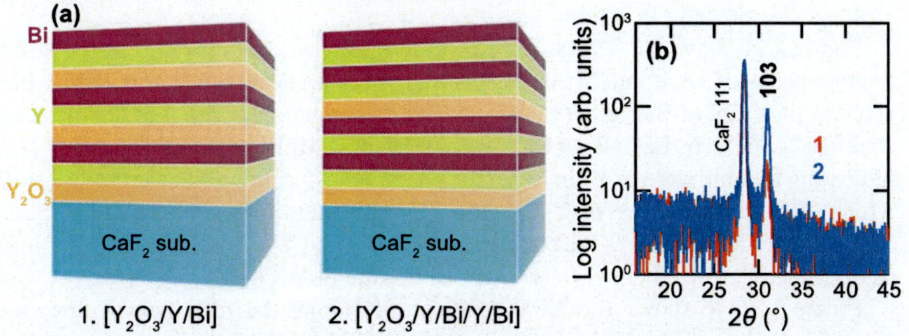

Fig. 3.10 a Schematic images of two types of multilayered precursors. b The comparison of XRD patterns of the Y_2O_2Bi epitaxial thin films fabricated from both types of multilayered precursors

later. The FWHM of 0.53 ° for the Y_2O_2Bi 006 rocking curve in the inset of Fig. 3.9a also indicated the high crystallinity. Four-fold symmetry in XRD in-plane ϕ scan of the Y_2O_2Bi 103 peak (Fig. 3.9b) indicated the epitaxial relationships as Y_2O_2Bi [100] || CaF_2 [110] and Y_2O_2Bi [001] || CaF_2 [001], consistent with the result in reductive SPE. The film surface roughness was 1.1 nm in root mean square at 2 mm square (Fig. 3.9c) without any surface residue. These results indicate the remarkable improvement of the Y_2O_2Bi epitaxial thin film than those with reductive SPE. Y_2O_2Bi epitaxial thin films were not obtained on oxide substrates such as $SrTiO_3$ and $LaAlO_3$ like reductive SPE.

From the comparison between two precursors with different sequences, [Y_2O_3/Y/Bi] multilayer and [Y_2O_3/Y/Bi/Y/Bi] multilayer, the latter showed larger XRD peak intensity (Fig. 3.10). This result suggests that the latter multilayer promoted an efficient formation of YBi alloy due to the larger number of Y/Bi interface in addition to high reaction efficiency of these multilayers compared with powder precursor. These factors were indispensable to achieve Y_2O_2Bi phase, enabling one-step heating procedure and highly crystalline quality.

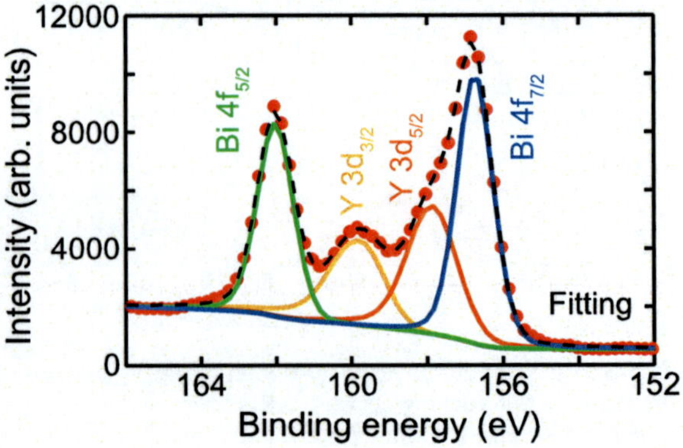

Fig. 3.11 XPS spectrum of the Y_2O_2Bi epitaxial thin film fitted by Bi 4f and Y 3d doublet peaks

Figure 3.11 shows an XPS spectrum of Y_2O_2Bi epitaxial thin film fitted by two doublet peaks of Bi 4f and Y 3d. The Bi $4f_{7/2}$ peak was located at 156.7 eV, which is lower than that of Bi metal (157.0 eV) [8] and comparable with that of Bi anionic state (Bi^{-}–Bi^{3-}) in $LaGaBi_2$ (156.6 eV) [9] and R_3MnBi_5 (156.7–156.8 eV) [10], indicating anionic state of Bi in Y_2O_2Bi. The Y $3d_{5/2}$ peak was located at 157.9 eV, which was between those of Y_2O_3 (156.7 eV) [11] and YF_3 (159.7 eV) [12], confirming Y^{3+} state as well as the coexistence of Y–O and Y–F bonds in $[Y_2O_2]^{2+}$ layer as explained below. These results verify Bi^{2-} state in Y_2O_2Bi.

Figure 3.12a–c shows the XPS depth profile along the film growth direction. Near the film surface, Bi loss due to the volatility under heating treatment was observed (Fig. 3.12a). The chemical composition was almost constant inside the film (Fig. 3.12b). In Fig. 3.12a,b, no peak shift was observed. At the film/substrate interface, all peaks were abruptly shifted about 2 eV to lower binding energy (93 nm in depth, Fig. 3.12c). This shift was attributed to the sample charge-up probably due to the too small volume of sputtered conducting Y_2O_2Bi film, because another Y_2O_2Bi film without ground by carbon tape showed a similar shift (Fig. 3.12d). Accordingly, XPS spectra measured in this study without that of excessively sputtered film were not influenced by the sample charge-up and the fluctuation of stoichiometry, hence these spectra indicated Bi^{2-} valence state.

It is noted that F was incidentally doped into the film due to the migration from the CaF_2 substrate during heating, confirmed by XPS measurement as shown in Fig. 3.13a. The amount of F in the film was higher for higher heating temperature. The *c*-axis length of Y_2O_2Bi film decreased with the presence of F (Fig. 3.13b), suggesting the substitution of F for O site due to the smaller ionic radius of F^- than that of O^{2-}, as was observed in structurally similar *R*FeAsO [13] and *R*OBiS$_2$ [14].

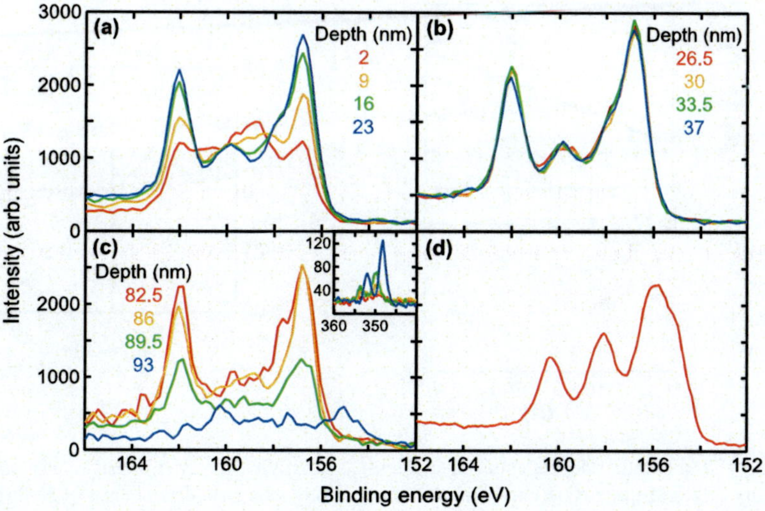

Fig. 3.12 XPS depth profile of Bi 4f and Y 3d peaks **a** near film surface, **b** inside the film, and **c** at the film/substrate interface. The inset of (**c**) shows Ca 2p spectra. **d** XPS spectrum of Bi 4f and Y 3d peaks measured without ground by carbon tape

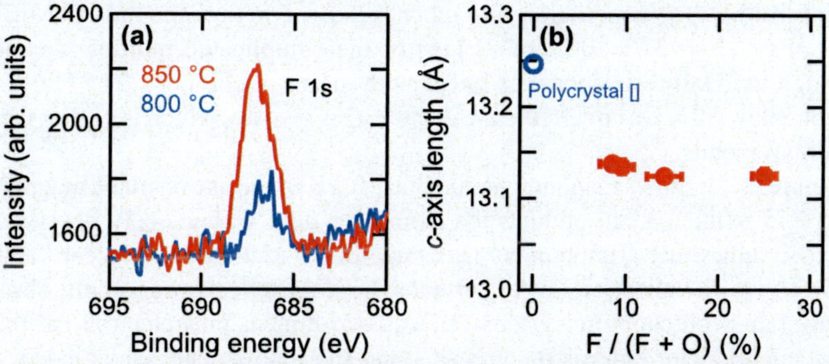

Fig. 3.13 **a** F 1 s spectra of the Y_2O_2Bi epitaxial thin films with different heating temperature. **b** F content dependence of *c*-axis length of the Y_2O_2Bi epitaxial thin films

3.3.3 Electronic Transport Properties

Figure 3.14a shows the ρ-T curve of Y_2O_2Bi epitaxial thin film. The film showed metallic behavior with one order smaller ρ than that of polycrystalline powder sample (10^3 μ Ω cm) [3] as a result of single crystallization of Y_2O_2Bi. The magnetic field dependence of Hall resistance (Fig. 3.14c) showed a negative slope at any temperature indicating the *n*-type conduction. This result is contradictory to the *p*-type conduction derived from the Seebeck measurement in the previous report [3]. The F content of 5×10^{21} cm^{-3} evaluated from XPS measurement was apparently smaller than that

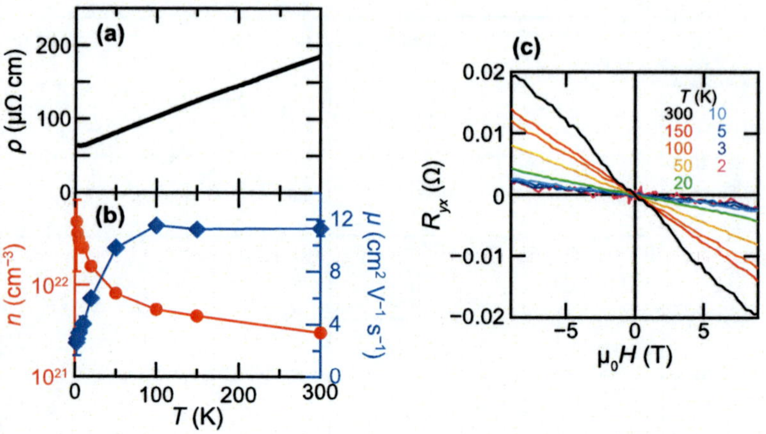

Fig. 3.14 Temperature dependences of **a** electrical resistivity, **b** carrier density (red), and carrier mobility (blue) of the Y_2O_2Bi epitaxial thin film. **c** Magnetic field dependence of Hall resistance (R_{yx}). R_{yx} was derived from $R_{yx} = [R_{yx}(+H) - R_{yx}(-H)]/2$ in order to eliminate R_{xx} component that is even function of H

of maximum electron carrier density 5×10^{22} cm^{-3} (Fig. 3.14b), suggesting either the intrinsic n-type conduction like other Bi square net compounds $CaMnBi_2$ [15] and $LaPd_{1-x}Bi_2$ (3.7×10^{22} cm^{-3}) [16] or the complicated multiband conduction although the Hall resistance was linear with the magnetic field. The Hall mobility except below 50 K was over 10 cm^2 V^{-1} s^{-1} (Fig. 3.14b), which is higher than those of normal metals.

Figure 3.15a shows magnetoconductance at 2 K in the out-of-plane magnetic field. The rapid reduction of conductance from zero field was scarcely observed in the other Bi square net compounds. This cusp-like variation is a typical behavior of the weak anti-localization (WAL) effect. The WAL effect was usually observed in strong spin–orbit coupling systems such as two-dimensional electron gas in narrow gap semiconductor heterostructure [17] and the topological insulator [18, 19]. In a two-dimensional electronic system, the Hikami–Larkin–Nagaoka (HLN) model can be applied to the WAL effect [20]. In the magnetic field range up to 9 T in this study, the magnetoconductance is described by the original HLN equation [20–22] expressed as

$$\Delta G(B) = -\alpha \frac{e^2}{2\pi^2\hbar}\left[2\eta\left(\frac{B_e}{B}\right) - 2\eta\left(\frac{B_{SO}^z + 2B_{SO}^x + B_\phi}{B}\right) + \eta\left(\frac{B_\phi}{B}\right) - \eta\left(\frac{4B_{SO}^x + B_\phi}{B}\right)\right] \tag{3.4}$$

where $\Delta G(B) = G(B) - G(0)$, α is the WAL coefficient equal to -0.5 in the ideal two-dimensional limit of a symplectic case such as surface state of topological insulator [18–20],[3] e is the elementary charge, $\hbar$ is Dirac's constant, η function is defined as $\eta(x) = \ln(x) - \psi((1/2) + x)$, ψ is the digamma function,

[3] The α equal to 1 or 0 corresponds to orthogonal or unitary case, respectively [20].

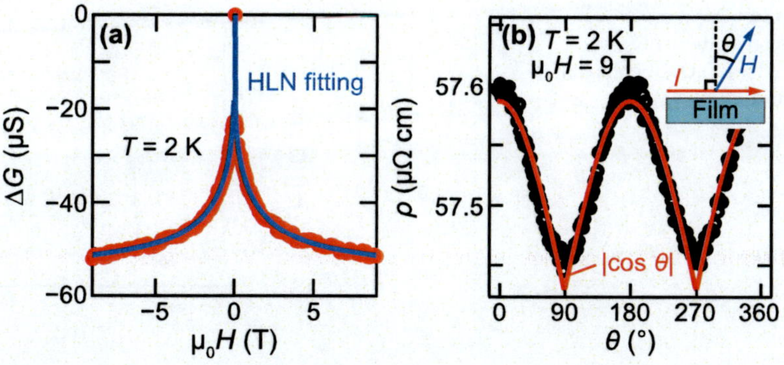

Fig. 3.15 **a** Magnetoconductance at 2 K superposed with fitting curve by the HLN model. **b** Angular dependence of MR at 2 K and 9 T. Inset shows the measurement configuration: *I* denotes current along the *ab*-plane, and θ denotes the tilt angle between the film normal and applied magnetic field *H*

$B_i = 1/4eL_i^2 (i = \mathrm{e, SO}, \phi)$, L_e is the elastic scattering length, L_SO is the spin–orbit scattering length along x or z direction, L_ϕ is the phase coherence length, and $B = \mu_0 H$ (μ_0: vacuum permeability). Fitting parameters were α and L_i. Here, the x and z directions correspond to the *a*- (or *b*-) and *c*-axis directions of Y_2O_2Bi, respectively [20, 22]. The blue line in Fig. 3.15a represents the fitting result of the HLN model. The obtained $\alpha = -0.49$ was close to the ideal two-dimensional limit of the symplectic case and similar to that of the bulk insulating topological insulator Bi_2Se_2Te (−0.56) [19] rather than that of the bulk conducting topological insulator Bi_2Te_3 (−0.39) [18]. The obtained scattering lengths were $L_\mathrm{SO}^x = 71.9\,\mathrm{nm}$, $L_\mathrm{SO}^z = 14.5\,\mathrm{nm}$, and $L_e = 31.3\,\mathrm{nm}$. The relations $L_\mathrm{SO}^x > L_\mathrm{SO}^z$ and $L_e > L_\mathrm{SO}^z$ indicated the two-dimensional nature and the strong spin–orbit coupling, respectively [20, 22]. The obtained L_ϕ =724 nm comparable to those of Bi_2Se_2Te and Bi_2Te_3 (318 and 331 nm, respectively) [18, 19] ensured the reasonable fitting result. These results indicated not only the two-dimensional nature but also strong spin–orbit coupling in Y_2O_2Bi originating from Bi^{2-} square net, similar to the topological insulator. The two-dimensional electronic transport was also confirmed in the angular dependence of magnetoresistance (MR). Figure 3.15b shows angular dependence of MR at 2 K and 9 T. MR was well-fitted to $|\cos\theta|$, indicating that MR varied by a field component perpendicular to the conducting channel [23, 24], i.e., Bi^{2-} square net. The small change in MR compared with $SrMnBi_2$ and $CaMnBi_2$ was probably due to the absence of the spin fluctuation [23, 25].

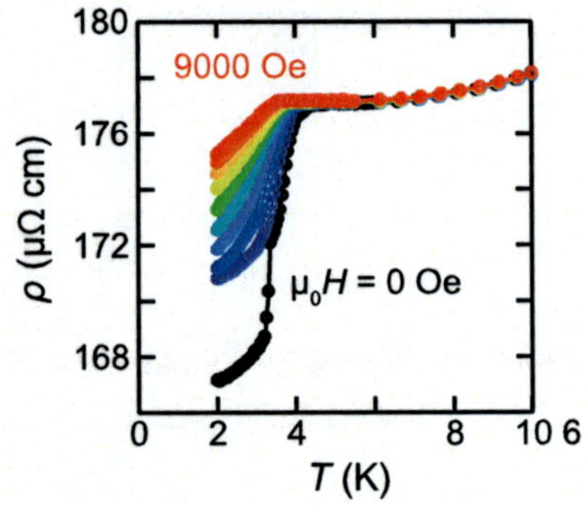

Fig. 3.16 Onset of superconductivity observed in a certain Y_2O_2Bi epitaxial thin film

3.3.4 *Onset of Superconductivity*

Figure 3.16 shows the ρ-T curves at low temperature and various fields of a certain Y_2O_2Bi epitaxial thin film. Notably, a sudden resistance-drop was observed around 4.0 K for the first time in this compound, and the transition was suppressed by applying the magnetic field, suggesting superconducting transition. The onset of superconductivity was observed in several samples with varied transition temperature T_c from 3 to 4 K. In various compounds composed of Bi, Y, and O elements, only a simple Bi shows superconductivity: Bi element ($T_c = 5.2\,\mu$ K) [26], high pressure phase (3.9 K for > 2.55 Pa, 7.2 K for >2.7 Pa, 8.3 K for >7.7 Pa) [27–30], nanowire (7.2 and 8.3 K) [30], granular film (4.3 K) [31], and amorphous phase (6.12 K) [32]. Observed T_c of Y_2O_2Bi was different from those in superconducting Bi, indicating intrinsic superconductivity in Y_2O_2Bi. However, Meissner effect and zero-resistance confirming bulk superconducting nature were not observed. Therefore, superconducting volume fraction seemed to be very small, and the driving force to induce superconductivity was unclear.

3.4 Conclusion

I succeeded in the fabrication of $ThCr_2Si_2$-type Y_2O_2Bi epitaxial thin film for the first time by developing two novel SPE methods: reductive SPE and multilayer SPE. In reductive SPE, Bi metal was reduced by Y metal without the formation of the C-rare earth type Y_2O_3 at the first heating, followed by the epitaxial growth of Y_2O_2Bi with the Bi^{2-} square net on the lattice-matched CaF_2 substrate at the second heating. This concept would be useful for the epitaxial growth of not only layered $ThCr_2Si_2$-type and analogous compounds but also other reductive compounds. However, the surface residue hampered the fundamental evaluation of physical properties.

Therefore, I improved reductive SPE by utilizing multilayered precursor. Crystalline quality of Y_2O_2Bi epitaxial thin film was dramatically enhanced and unusual valent of Bi^{2-} state was directly observcd. This multilayer SPE was universally applicable to epitaxial growth of other R_2O_2Bi such as R_2O_2 [33] and Dy_2O_2Bi (Appendix D). Magnetotransport measurements indicated the two-dimensional electronic nature in addition to strong spin–orbit coupling in the Bi^{2-} square net, reminiscent of the topological insulator. Onset of superconducting transition was observed around 4 K although superconducting volume fraction was quite small. R_2O_2Bi is expected to be a good platform to investigate the exotic two-dimensional electronic system with the strong spin–orbit coupling.

References

1. Hiramatsu H, Ueda K, Ohta H, Orita M, Hirano M, Hosono H (2002) Appl Phys Lett 81(4):598
2. Nomura K, Ohta H, Ueda K, Kamiya T, Hirano M, Hosono H (2003) Science 300(5623):1269
3. Mizoguchi H, Hosono H (2011) J Am Chem Soc 133(8):2394
4. Massalski TB, Okamoto H, Subramanian PR, Kacrpzak L (eds) Binary alloy phase diagrams 2nd ed. ASM International
5. Nuss J, Jansen M (2009) J Alloy Compd 480(1):57
6. Nuss J, Jansen M (2012) Z Anorg Allg Chem 638(3–4):611
7. Wang PL, Kolodiazhnyi T, Yao J, Mozharivskyj Y (2012) J Am Chem Soc 134(3):1426
8. Moulder JF, Stickle WF, Sobol PE, Bomben KD (1995) Handbook of x-ray photoelectron spectroscopy a reference book of standard spectra for identification and interpretation of XPS data. Physical Electronics, Inc
9. Morgan MG, Wang M, Chan WY, Mar A (2003) Inorg Chem 42(5):1549
10. Zelinska OY, Mar A (2008) Inorg Chem 47(1):297
11. de Rouffignac P, Park JS, Gordon RG (2005) Chem Mater 17(19):4808
12. Wang M, Huang QL, Zhong HX, Chen XT, Xue ZL, You XZ (2007) Cryst Growth Des 7(10):2106
13. Kamihara Y, Watanabe T, Hirano M, Hosono H (2008) J Am Chem Soc 130(11):3296
14. Mizuguchi Y, Demura S, Deguchi K, Takano Y, Fujihisa H, Gotoh Y, Izawa H, Miura O (2012) J Phys Soc Jpn 81(11):114725
15. Wang K, Wang L, Petrovic C (2012) Appl Phys Lett 100(11):112111
16. Han F, Malliakas CD, Stoumpos CC, Sturza M, Claus H, Chung DY, Kanatzidis MG (2013) Phys Rev B 88(14):144511
17. Koga T, Nitta J, Akazaki T, Takayanagi H (2002) Phys Rev Lett 89(4):046801
18. He HT, Wang G, Zhang T, Sou IK, Wong GK, Wang JN, Lu HZ, Shen SQ, Zhang FC (2011) Phys Rev Lett 106(16):166805
19. Bao L, He L, Meyer N, Kou X, Zhang P, Chen ZG, Fedorov AV, Zou J, Riedemann TM, Lograsso TA et al (2012) Sci Rep 2
20. Hikami S, Larkin AI, Nagaoka Y (1980) Progress Theoret Phys 63(2):707
21. Zhang S, McDonald R, Shekhter A, Bi Z, Li Y, Jia Q, Picraux ST (2012) Appl Phys Lett 101(20):202403
22. Dey R, Pramanik T, Roy A, Rai A, Guchhait S, Sonde S, Movva HC, Colombo L, Register LF, Banerjee SK (2014) Appl Phys Lett 104(22):223111
23. Wang K, Graf D, Lei H, Tozer S, Petrovic C et al (2011) Phys Rev B 84(22):220401
24. Tang H, Liang D, Qiu RL, Gao XP (2011) ACS Nano 5(9):7510
25. He J, Wang D, Chen G (2012) Appl Phys Lett 100(11):112405
26. Prakash O, Kumar A, Thamizhavel A, Ramakrishnan S (2016) Science:aaf8227 (2016)
27. Stromberg HD, Stephens DR (1964) J Phys Chem Solids 25:1015
28. Brandt B, Ginzburg NI (1969) Contemp Phys 10:355

29. Wittig J (1966) Zeitschrift fuer Physik 195:215
30. Tian M, Wang J, Kumar N, Han T, Kobayashi Y, Liu Y, Mallouk TE, Chan MH (2006) Nano Lett 6(12):2773
31. Vossloh C, Holdenried M, Micklitz H (1998) Phys Rev B 58(18):12422
32. Watanabe K, Noto K, Toyota N, Muto Y (1984) J Phys Soc Jpn 53(4):1444
33. Shibata S, Sei R, FUkumura T, Hasegawa T (submitted)

Chapter 4
Two-Dimensional Superconductivity in Polycrystalline Y_2O_2Bi

4.1 Introduction

Superconducting Bi square net would provide a new platform to explore the interaction between low-dimensional superconductivity and strong spin–orbit coupling such as topological superconductivity [1], owing to the possibility of a two-dimensional topological insulator [2]. So far, $AMBi_2$ compounds with Bi^- square net have been extensively studied and several compounds were reported to be superconducting [3–5]. However, superconductivity was often subject to small superconducting volume fraction or was attributed to neighboring layer rather than Bi^- square net (see Sect. 1.2.4 and Table 4.1), indicating little hope to achieve superconducting Bi square net in $AMBi_2$ compounds.

$ThCr_2Si_2$-type layered R_2O_2Bi (R: rare earth) is a new series of compounds composed of conducting Bi^{2-} square net and insulating $[R_2O_2]^{2+}$ layer [6]. This compound can be regarded as a counterpart of iso-structural FeAs-based high-temperature superconductor $BaFe_2As_2$ [7] with their inverted carrier conduction paths, Bi^{2-} square net and $[Fe_2As_2]^{2+}$ layer, respectively (Fig. 4.1), possibly connecting to the emergence of superconductivity. However, R_2O_2Bi has been already concluded as non-superconducting [8], and the application of chemical pressure just changes the electrical transport properties from insulating in La_2O_2Bi to metallic in Er_2O_2Bi [6]. In addition, quite large carrier density ($\geq 10^{22}$ cm^{-3}, Sect. 3.3.3), which is consistent with one anti-bonding electron per one Bi^{2-} anion (Sect. 1.2.1), demonstrates no effect of further carrier doping, indicating the conventional approach was not applicable to induce superconductivity in R_2O_2Bi.

In iso-structural $BaFe_2As_2$, decreasing the inter-layer distance by applying uniaxial pressure along c-axis direction weakens the superconducting properties, similar to that observed in cuprate high-temperature superconductor [9, 10]. This is consistent

This section contains the contents of the following publication. Reprinted with permission from R. Sei, S. Kitani, T. Fukumura, H. Kawaji, T. Hasegawa, *Journals of the American Chemical Society* **138**, 11085 (2016). Copyright 2016 by American Chemical Society.

R. Sei, *Two-Dimensional Superconductivity in Rare Earth Oxybismuthides with Unusual Valent Bismuth Square Net*, Springer Theses,
https://doi.org/10.1007/978-981-99-7313-2_4

Table 4.1 Properties of Bi square net compounds. SC: superconducting

Compounds	T_c (K)	Crystal structure (space group)	Principle SC component	Notes	References
$CeNi_xBi_2$	4.2	$ZrCuSi_2$ (*P4/nmm*)	Bi^- square net [3] or $NiBi_3$ impurity [11–13] (under debate)	Small SC volume fraction (< 17%) except for ref. [3]. No Ni deficiency dependence of T_c, irreproducible superconductivity, and the same T_c with that of $NiBi_3$	[3, 11–13]
$LaPd_xBi_2$	2.1	$ZrCuSi_2$ (*P4/nmm*)	Pd	Effect of Pd vacancy on Fermi surface and CDW	[4, 14]
Bi/$SrMnBi_2$	5.7		Strained Bi	Blob-like Bi on $SrMnBi_2$ single crystal showed superconductivity due to the strain. Large sample dependence	[5]
RENi_xBi_2 (*RE*: Y, La, Nd)	~4	$ZrCuSi_2$ (*P4/nmm*)	Bi^- square net [3] or $NiBi_3$ impurity [11] (under debate)	Small SC volume fraction (1–17%) and similar T_c with that of $NiBi_3$	[3, 11]
RENi_xBi_2 (*RE*: Pr, Nd, Sm, Gd, Tb, Dy)	–	$ZrCuSi_2$ (*P4/nmm*)	–	AF transition at 3–10 K	[11]
$CePd_xBi_2$	–	$ZrCuSi_2$ (*P4/nmm*)	–	AF transition at 6 K	[4]
$SrMnBi_2$	–	$SrZnBi_2$ (*I4/mmm*)	–	AF transition at 290 K. Anisotropic Dirac fermion	[15, 16]
$CaMnBi_2$	–	$ZrCuSi_2$ (*P4/nmm*)	–	AF transition at 250 K. Anisotropic Dirac fermion	[17, 18]

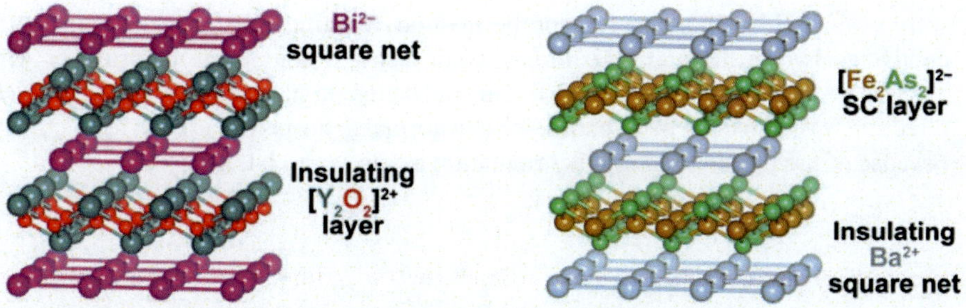

Fig. 4.1 The crystal structures of $ThCr_2Si_2$-type Y_2O_2Bi (left) and $BaFe_2As_2$ (right). SC: superconducting

with the very small superconducting volume in F doped Y_2O_2Bi epitaxial thin film with c-axis shrinking (Sect. 3.3.4). Therefore, the inverted effect, i.e., expanding c-axis length would be one alternative route to induce superconductivity in Bi^{2-} square net of R_2O_2Bi. In this study, I discovered the two-dimensional superconductivity of Bi^{2-} square net in polycrystalline Y_2O_2Bi at about 2 K via chemical control of the structure. It is noted that the superconductivity was not observed in pristine Y_2O_2Bi but induced by expanding Bi^{2-} inter-net distance through excessive O incorporation into the hidden interstitial site. This result indicates an important role of enhanced two-dimensionality of Bi^{2-} square net in the emergence of the superconductivity, and provides a new chemical approach to explore the superconductivity in layered compounds.

4.2 Experimental

Polycrystalline Y_2O_2Bi samples were synthesized as follows. The starting materials were Bi (3N), Y (3N), and Y_2O_3 (3N) powders. The Y_2O_3 powder was heated at 350°C for 1 day before use in order to eliminate absorbed moisture. Nominal amount of $Y_2O_xBi_{1.5}$ $(1.1 \leq x \leq 2.0)$ mixed powders were pressed into pellets by ~ 5 MPa and covered with Ta foil. The excess amount of Bi was added in order to compensate volatile Bi and remove burnable Y residue in air by forming YBi, and Bi residue evaporated from sample pellets[1]. This operation was performed in an Ar-filled glovebox. The pellets were heated in evacuated quartz tubes at 500°C for 7.5 h, followed by heating at 1000°C for 20 h. The products were ground in the glovebox, pressed into pellets in air by ~ 30 MPa, covered with Ta foil, and heated in evacuated quartz tubes at 1000°C for 10 h. Crystal structures were evaluated by powder XRD using Cu Kα radiation. Rietveld refinements were performed by RIETAN-FP [19]. Magnetic properties were measured by SQUID magnetometer. Electrical transport properties

[1] The nominal amount of Bi was changed to 1.8, 1.7, and 1.6 for $x = 1.1$, 1.2, and 1.3, respectively, in order to compensate increased Y residue.

were evaluated by standard four-probe method. Specific heat measurements were conducted by the thermal relaxation method using a homemade calorimeter with a $^3He/^4He$ dilution refrigerator [20]. Out-of-plane and in-plane magnetic field (H) direction was defined to be H parallel to and normal to the pellet surface, respectively, because of the preferential c-axis orientation as discussed later.

4.3 Crystal Structure

Figure 4.2 shows powder XRD patterns of all Y_2O_2Bi samples ($1.1 \le x \le 2.0$). Refined structural parameters are summarized in Table 4.2. Most of the diffraction peaks were completely fitted by simulation as $ThCr_2Si_2$-type Y_2O_2Bi phase (Fig. 4.3a), while several non-superconducting minor phases appeared depending on the prescribed amount of O: YBi for reductive condition ($1.1 \le x \le 1.4$) and Y_2O_3 for oxidative condition ($1.5 \le x \le 2.0$) (Fig. 4.3b). The typical atomic ratio of Y/Bi in Y_2O_2Bi phase was almost stoichiometric, e.g., 2.02 for both reductive ($x = 1.2$) and oxidative ($x = 1.6$) conditions, evaluated by inductively coupled plasma mass spectroscopy and Rietveld refinement. The lattice constants for reduc-

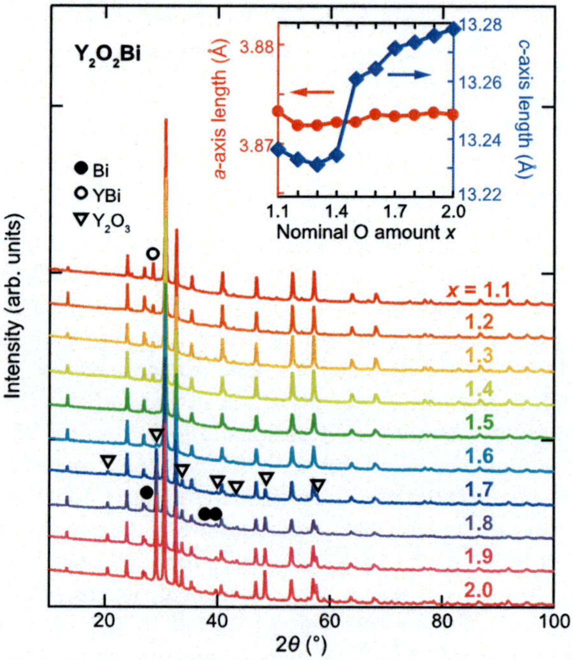

Fig. 4.2 XRD patterns of Y_2O_2Bi samples. Inset shows lattice constant of each sample. Results of the Rietveld refinements are seen in Table 4.2 and Fig. 4.3

Table 4.2 Summary of structural data in Y_2O_2Bi samples

x	1.1	1.2	1.3	1.4	1.5	1.6	1.7	1.8	1.9	2.0
Space group	*I4/mmm*	*I4/mmm*	*I4/mmm*	*I4/mmm*	*I4/mmm*	*I4/mmm*	*I4/mmm*	*I4/mmm*	*I4/mmm*	*I4/mmm*
a (Å)	3.87327(11)	3.87183(12)	3.87183(17)	3.87215(11)	3.87217(18)	3.8730(2)	3.87278(14)	3.87329(2)	3.87317(14)	3.87297(13)
c (Å)	13.2363(4)	13.2329(4)	13.2311(6)	13.2345(4)	13.2607(6)	13.2643(7)	13.2713(5)	13.2734(8)	13.2758(5)	13.2780(5)
V (Å^3)	198.573(10)	198.375(11)	198.348(15)	198.431(10)	198.828(16)	198.964(18)	199.050(13)	199.095(19)	199.156(12)	199.168(12)
Y z	0.33266(14)	0.33336(14)	0.33309(15)	0.33222(11)	0.33268(15)	0.33187(13)	0.33164(13)	0.3326(2)	0.33143(13)	0.33168(15)
Purity (mol%)	91.9	85.1	88.4	96.9	98.1	96.3	70.9	57.7	58.5	46.8
R_{wp}	2.844	2.828	3.298	2.600	2.595	2.875	1.999	2.406	2.078	2.238
R_e	1.587	1.592	1.590	1.626	1.609	1.614	1.579	1.574	1.577	1.581
S	1.792	1.776	2.074	1.599	1.613	1.781	1.266	1.528	1.318	1.416

Wyckoff positions: Bi on the 2*a* site (0, 0, 0), Y on the 4*e* site (0, 0, z), and O on the 4*d* site (0, 0.5, 0.25)

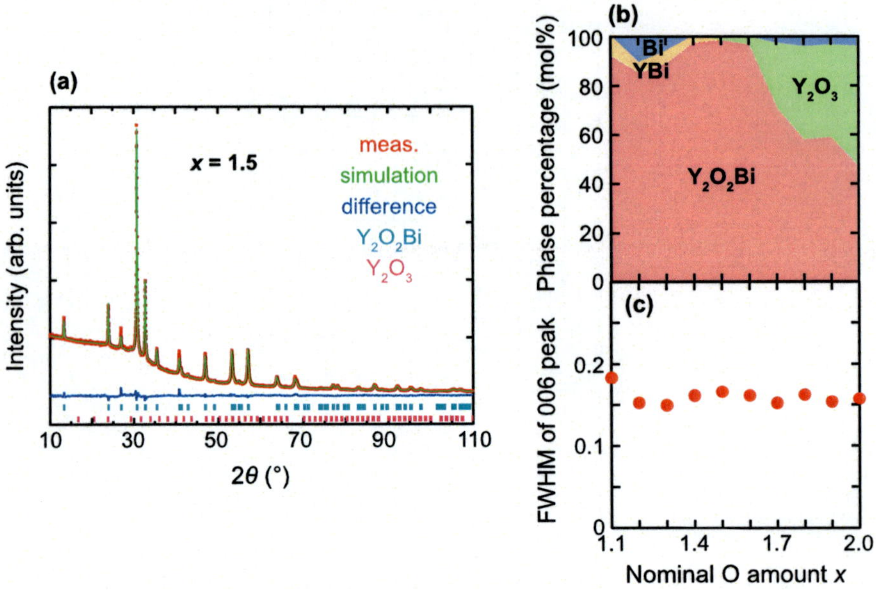

Fig. 4.3 **a** Typical result of Rietveld refinement for $x = 1.5$. **b** Phase diagram for Y_2O_2Bi samples. **c** The FWHM of 006 peaks in 2θ-θ patterns for Y_2O_2Bi samples

tive samples were similar to those of Y_2O_2Bi polycrystalline powder ($a = 3.8734$, $c = 13.2469$ Å) [6], indicating their ideal composition (inset of Fig. 4.2). On the other hand, the c-axis showed abrupt expansion for $x = 1.5$ followed by a gradual increase with almost constant a-axis toward a $x = 2.0$ (inset of Fig. 4.2). This expansion was possibly caused by O incorporation between Bi^{2-} square net and adjacent Y termination layers implied by the jump of those distances between samples $x = 1.4$ and $x = 1.5$ in contrast with continuously changed distance between Y termination layer and adjacent O layer (Fig. 4.4). The FWHM of 006 diffraction peaks in 2θ-θ patterns showed similar values among all Y_2O_2Bi samples, indicating that O was homogeneously incorporated into Y_2O_2Bi framework in each sample without the deterioration of the crystalline quality (Fig. 4.3c). Thus, I succeeded in introducing the chemical control of the structure by varying the amount of O in Y_2O_2Bi. This incorporation process into the hidden interstitial site was essentially different from the conventional intercalation technique into van del Waals gap: the carrier doping was critical for tuning physical properties in the latter case [21–24], while the structural control was essential in the former case as discussed later.

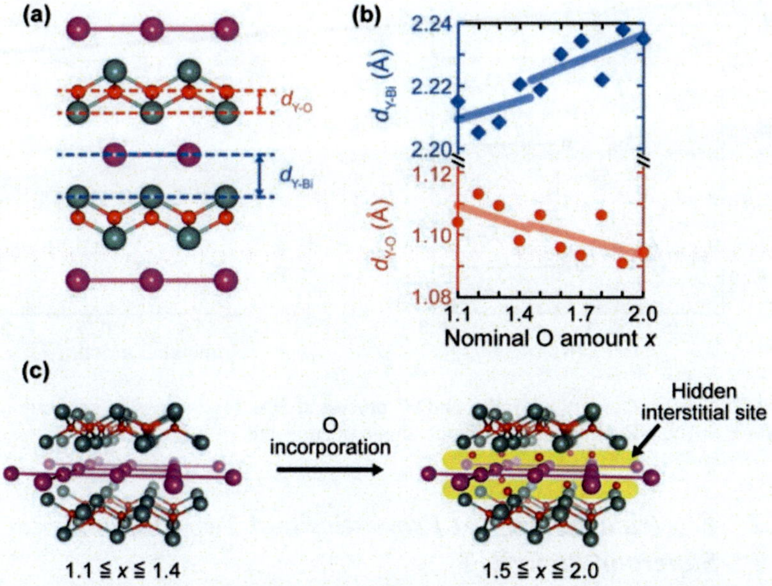

Fig. 4.4 **a** Crystal structure of Y_2O_2Bi along [100] direction. The $d_{Y\text{-}O}$ ($d_{Y\text{-}Bi}$) denotes the distance between Y termination layer and adjacent O (Bi) layer. **b** The sample dependence of $d_{Y\text{-}O}$ and $d_{Y\text{-}Bi}$. Thick lines correspond to the least square fitting lines. **c** Schematic image of the local crystal structure around Bi^{2-} square net without (left) and with (right) O incorporation into Wyckoff position 8g site (0, 0.5, 0.09)

4.4 Emergence of Bulk Superconductivity

4.4.1 Magnetic Properties

Figure 4.5a shows the χ-T curves in zero-field cooling (ZFC) and field cooling (FC) at 5 Oe. No diamagnetic signal was observed for reductive samples ($1.1 \leq x \leq 1.4$), whereas large diamagnetic signals due to the Meissner effect were observed around 2 K for oxidative samples ($1.5 \leq x \leq 2.0$) corresponding to the superconducting transition. The critical temperature T_c was a monotonically increasing function of nominal amount of O, i.e., x. The magnetization curve at 2 K for $x = 1.9$ indicated the type-II superconductivity with a lower critical magnetic field of 18 Oe deduced from the magnetization minimum (inset of Fig. 4.5a). Shielding volume fractions calculated from the magnetization curves for oxidative samples confirmed the bulk superconductivity of Y_2O_2Bi (Fig. 4.5b). The shielding volume fraction over 100% was caused by the polycrystalline nature of the samples.

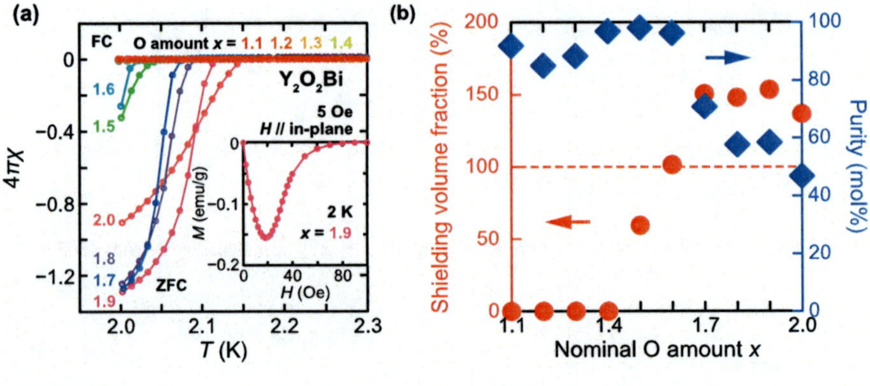

Fig. 4.5 **a** The χ-T curves with ZFC and FC process at 5Oe. Inset shows magnetization curve at 2 K for $x = 1.9$. **b** Evolution of shielding volume fraction and Y_2O_2Bi phase purity

4.4.2 Electrical Transport Properties and Two-Dimensional Superconductivity

Hereafter, we mainly discuss the properties of $x = 1.6$ because of the highest purity of Y_2O_2Bi phase (96.3 mol%) and the bulk superconductivity with sufficiently high shielding volume fraction (Fig. 4.5b). Figure 4.6a shows the ρ-T curves at various magnetic fields. The sample showed metallic behavior (inset of Fig. 4.6a), similar to that of the previous report [6]. The onset of superconductivity and zero-resistance at 0 Oe were observed at 2.04 K and at 2.01 K, respectively, and the T_c decreased with increasing magnetic field. The superconducting parameters of T_c, upper critical field H_{c2} (.$T = 0$ K), and coherence length ξ ($H_{c2} = \Phi_0/2\pi\xi^2$; Φ_0: the flux quantum) changed monotonically for oxidative samples as observed in magnetic measurement

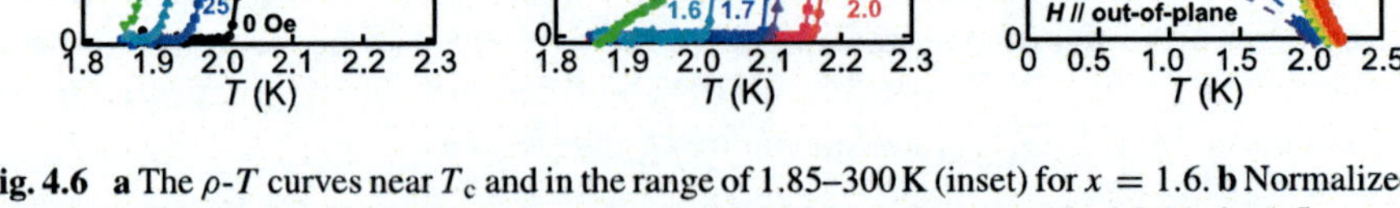

Fig. 4.6 **a** The ρ-T curves near T_c and in the range of 1.85–300 K (inset) for $x = 1.6$. **b** Normalized ρ-T curves near T_c for all samples. **c** Temperature dependence of upper critical fields for $1.5 \le x \le 2.0$. Inset shows magnified view near T_c. Broken lines denote fitting result with Ginzburg–Landau theory $H_{c2}(T) = H_{c2}(0)[1 - (T/T_c)^2]$

Table 4.3 Superconducting parameters for $1.5 \leq x \leq 2.0$

x	1.5	1.6	1.7	1.8	1.9	2.0
T_c (K)	2.02	2.04	2.12	2.14	2.19	2.21
H_{c2} (Oe)	574	841	1338	1382	1559	1864
ξ (nm)	75.8	62.6	49.6	48.8	46.0	42.0

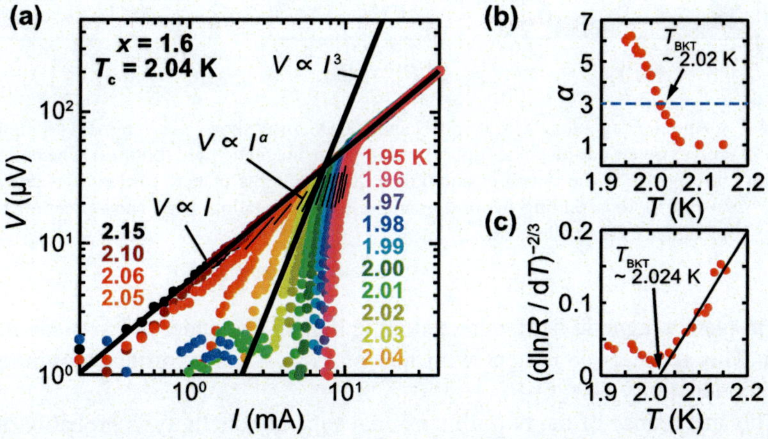

Fig. 4.7 **a** V-I curves at various temperatures near T_c. Black lines denote the fitting result with $V \propto I^{\alpha}$ scaling law. **b** Temperature dependence of the exponent α. **c** Temperature dependence of $(d \ln R/dT)^{-2/3}$

from 2.02 K, 574 Oe, and 75.8 nm to 2.21 K, 1864 Oe, and 42.0 nm, respectively, as summarized in Fig. 4.6b, c and Table 4.3.

Electrical transport properties in Y_2O_2Bi exhibited two-dimensional superconductivity, supporting the superconducting Bi^{2-} square net. Figure 4.7a shows voltage–current characteristics (V-I curves) near critical current around T_c. The V-I curves showed a transition of scaling law $V \propto I^{\alpha}$ above T_c from $\alpha = 1$ (ohmic law) to 3 and more with decreasing temperature. This tendency corresponded to the BKT transition at $T_{BKT} = 2.02$ K as seen in Fig. 4.7b [25, 26]. This value was in good coincidence with T_{BKT} independently derived from the ρ-T curve according to the relationship $R(T) \propto \exp[-b(T/T_{BKT} - 1)^{-1/2}]$, in which b is the material constant (Fig. 4.7c) [27]. These results confirmed the two-dimensional superconductivity of Y_2O_2Bi, indicating the weak coupling between superconducting layers, i.e., Bi^{2-} square nets, as observed in cuprate [28, 29] and Fe-based [30, 31] superconductors.

The two-dimensionality was also manifested in angular dependent magnetoresistance as shown in Fig. 4.8a, indicating higher H_{c2} under an in-plane magnetic field. Irreproducible ρ at $\theta = 0°$ between forward and backward sweeps might be attributed to a decreased effective field due to the flux pinning effect in insulating $[Y_2O_2]^{2+}$ layer and/or grain boundary. Figure 4.8b shows the ρ-T curves under out-of-plane

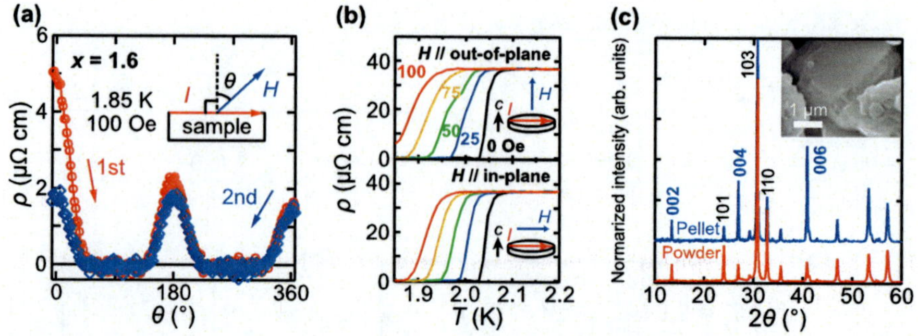

Fig. 4.8 **a** Angular dependent MR at 1.85 K and 100 Oe. Inset shows measurement configuration. **b** The ρ-T curves under out-of-plane (upper) and in-plane magnetic field (bottom). The pellet surface normal corresponds to the c-axis orientation. **c** XRD patterns of pellet (blue) and powder (red) forms for $x = 1.6$, representing the preferential c-axis orientation. Inset shows scanning electron microscopy image of the pellet

and in-plane magnetic field, demonstrating higher H_{c2} under the in-plane magnetic field. This result is consistent with that of Fig. 4.8a, supporting two-dimensional superconductivity of Y_2O_2Bi.

The appearance of the two-dimensional superconductivity is probably owing to the sufficiently high crystalline orientation of a large-sized crystal. Figure 4.8c shows XRD patterns of powder and pellet forms for $x = 1.6$. The Y_2O_2Bi $00l$ diffraction peaks were significantly enhanced in the pellet form, indicating the dominant c-axis orientation of Y_2O_2Bi crystal normal to the pellet surface. The apparently larger crystalline size than coherence length was confirmed by scanning electron microscopy (inset of Fig. 4.8c).

4.4.3 Specific Heat Measurement

Figure 4.9a shows the specific heat plotted as C/T versus T^2. A large jump in the specific heat was observed at T_c for each magnetic field, evidencing again the bulk superconductivity. An upturn below about 0.6 K was due to a Schottky anomaly originating from ^{209}Bi nuclei ($I = 9/2$). From the curve fitting of the normal state at 1000 Oe with the following equation, $C = \gamma_N T + \beta T^3 + \delta T^5 + AT^{-2}$, where $\gamma_N T$, $\beta T^3 + \delta T^5$, and AT^{-2} correspond to the electronic, phonon, and nuclear contributions, respectively; the Sommerfeld constant $\gamma_N = 2.71$ mJ mol^{-1} K^{-2}, $\beta = 0.582$ mJ mol^{-1} K^{-4}, $\delta = 0.0536$ mJ mol^{-1} K^{-6}, and $A = 0.325$ mJ mol^{-1} K were obtained. From the difference in the electronic specific heat C_{el} between superconducting and normal states, the specific heat jump at T_c was evaluated to be $\Delta C_{el}/\gamma_N T_c = 1.68$ (Fig. 4.9b), which was comparable to the Bardeen–Cooper–Schrieffer weak coupling limit 1.43.

The fitting parameters obtained for the specific heat data lead to important insights. The Sommerfeld constant γ_N is directly related to the density of states at the Fermi

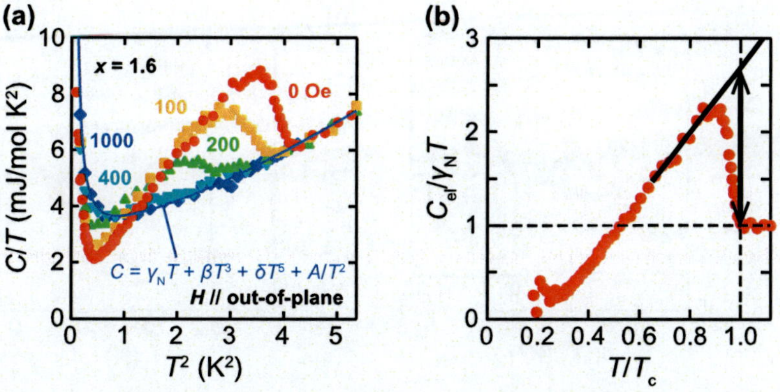

Fig. 4.9 **a** The specific heat plotted as C/T versus T^2 under various magnetic fields for $x = 1.6$. **b** Electronic contribution of specific heat C_{el} divided by $\gamma_N T$ as a function of normalized temperature T/T_c, where specific heat jump was evaluated to be $\Delta C_{el}/\gamma_N T_c = 1.68$ (vertical arrow). T_c was derived from the ρ-T curve

level, $D(\varepsilon_F)$, $\gamma = \pi 2k_B^2 D(\varepsilon_F)/3$, where k_B is the Boltzmann constant. The value of $D(\varepsilon_F)$ was obtained as 3.4×10^{21} eV^{-1} cm^{-3}, which is quite larger than that of bismuth metal, 3×10^{16} eV^{-1} cm^{-3} [32], but smaller than that of normal metals, such as Li metal, 33×10^{21} eV^{-1} cm^{-3} [33]. The Debye temperature Θ_D was calculated to be 256 K using the relationship $\beta = 12\pi^4 Rr/5\Theta_D^3$, where R is the gas constant and r is the number of atoms in the chemical formula. Assuming that Y_2O_2Bi is a BCS-type superconductor, the electron–phonon coupling constant λ can be calculated from McMillan's formula [34]:

$$T_c = \left(\frac{\Theta_D}{1.45}\right) \exp\left(-\frac{1.04(1+\lambda)}{\lambda - \mu^*(1+0.62\lambda)}\right) \tag{4.1}$$

where μ^* is the Coulomb pseudo-potential. Supposing the value of $\mu^* = 0.13$, which is a typical value for the low T_c metals, the substitution of the experimental values of Θ_D =256 K and T_c =2.04 K leads to λ =0.53. The obtained λ is in the weak coupling regime, and hence supports the analysis of the specific heat jump. The nuclear term AT^{-2} arises from the hyperfine splitting of the nuclear levels of ^{209}Bi ($I = 9/2$) due to the interaction between the nuclear quadrupole moment and the electric field gradient. Thus, the coefficient A can lead to the quadrupole coupling constant, e^2qQ, through the following equation [32]:

$$A = \frac{R}{80}\frac{(2I+2)(2I+3)}{2I(2I-1)}\left(\frac{e^2qQ}{kT}\right)^2. \tag{4.2}$$

The substitution results in $e^2qQ/k = 4.13 \times 10^{-2}$ K or, in terms of the frequency, 826 MHz, which is reasonable for bismuth compounds [35–37].

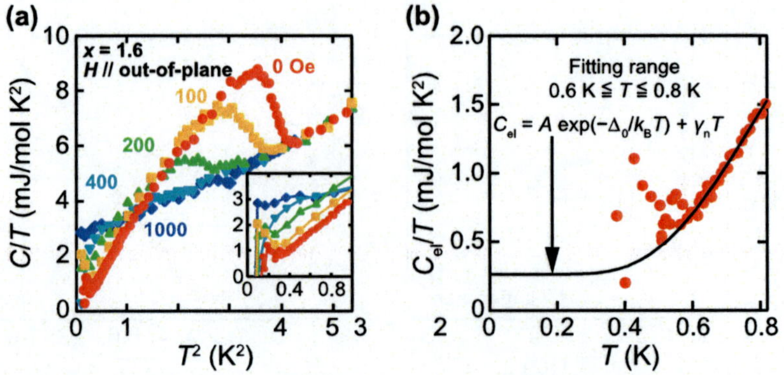

Fig. 4.10 **a** T^2 dependence of C/T after subtracting nuclear specific heat. **b** Temperature dependence of electronic specific heat $C_{\rm el}$ at very low temperature. Black line denotes the fitting curve assuming full-gap model in the range of $0.6\,{\rm K} \leq T \leq 0.8\,{\rm K}$

The estimated thermal effective mass was $m^*_{\rm th} \approx 0.6m_{\rm e}$ calculated from the relationship $m^*_{\rm th} = m_{\rm e}\gamma_{\rm N}/\gamma_{\rm e}$, where $m_{\rm e}$ and $\gamma_{\rm e}$ are the mass and Sommerfeld constant of free-electron, respectively, indicating light conducting carriers in Bi^{2-} square net. The $\gamma_{\rm e}$ was calculated from the relationship $\gamma_{\rm e} = m_{\rm e}(N_{\rm e}/\hbar^2)(\pi/3n)^{2/3}k_{\rm B}^2$, where $N_{\rm e}$ is the number of electrons, $\hbar$ is Dirac's constant, and n is the density of electrons. Here, we adopted $N_{\rm e} = 6 \times 10^{24}\,{\rm mol}^{-1}$, which was the number of carriers per 1 mole of Y_2O_2Bi estimated from the carrier density discussed later.

The specific heat data was applied to analyze the superconducting gap structure. The field dependence of the electronic specific heat coefficient $\gamma(H)$ which was derived from the intercept of C/T plotted by T^2, shows $\gamma(H) \propto H$ and $H^{1/2}$ dependence in full-gap and nodal superconductors, respectively [38]. In order to evaluate the relationship, the nuclear contribution AT^{-2} was subtracted from the C/T (Fig. 4.10a). The C/T became negative at the lowest temperature (inset of Fig. 4.10a), indicating the difficulty of quantitative discussion. Figure 4.10b shows preliminary analysis of electronic specific heat $C_{\rm el}$. In general, full-gap superconductor shows the following $C_{\rm el}(T)$ dependence at $T \ll T_{\rm c}$,

$$C_{\rm el}(T) = A\exp\left(-\frac{\Delta_0}{k_{\rm B}T}\right) + \gamma_{\rm n}T \tag{4.3}$$

where Δ_0 is the superconducting gap and $\gamma_{\rm n}T$ is the contribution from the non-superconducting phase such as impurities. We performed the above fitting in the range of $0.6\,{\rm K} \leq T \leq 0.8\,{\rm K}$, resulting in $\Delta_0 = 0.26\,{\rm meV}$ and $2\Delta_0/k_{\rm B}T_{\rm c} = 3.00$, where the latter was smaller than the BCS limit of 3.54. The superconducting volume fraction derived from $1 - \gamma_{\rm n}/\gamma_{\rm N}$ was 90.3%, showing good coincidence with the phase purity 96.3mol% (Fig. 4.3b).

4.5 Mechanism of Superconductivity

Carrier doping is a powerful technique to induce superconductivity. However, a similar shape of ρ-T curves and RRR irrespective of the samples ruled out the possibility of carrier doping as an effect of O incorporation. Figure 4.11a,b shows ρ-T curves and T_c versus RRR plot for non-superconducting ($1.1 \leq x \leq 1.4$) and superconducting ($1.5 \leq x \leq 2.0$) samples. Samples for $1.1 \leq x \leq 1.6$ showed similar ρ and RRR values. In addition, the ρ-T curves of all samples showed very similar shape despite more than two decades of change in ρ value. These results represent that the carrier density was not significantly changed for all samples, suggesting that the O incorporation did not cause the carrier doping. The decrease in conductivity and RRR for $1.7 \leq x \leq 2.0$ with increased nominal amount of O was mainly attributed to decreased Y_2O_2Bi volume (Fig. 4.3b), and increased inter-net distance resulted in decreased inter-net conductance.

Indeed, carrier density showed almost constant values between non-superconducting and superconducting samples. Figure 4.11c shows the carrier density of high-purity samples ($1.3 \leq x \leq 1.6$) derived from Hall resistance versus magnetic field at 100 K. These samples showed p-type conduction with rather constant carrier density ($\geq 10^{22}$ cm^{-3}) without the systematic dependence of O con-

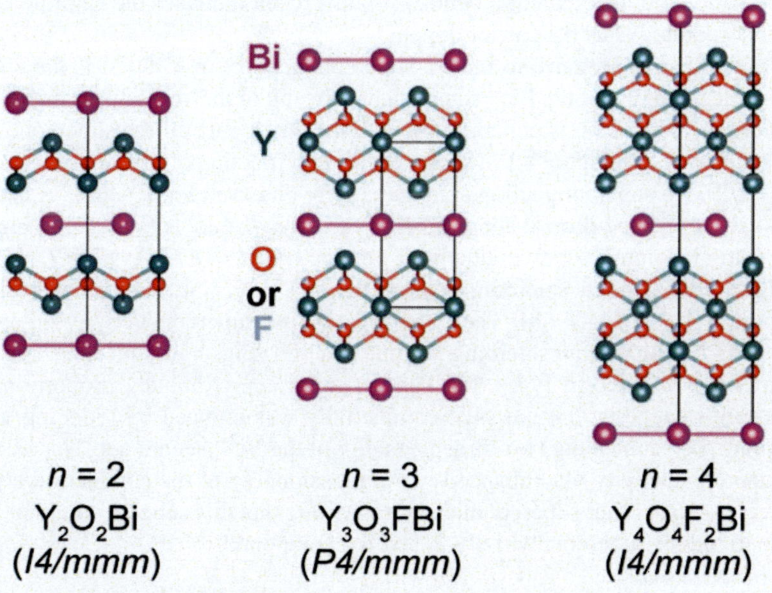

Fig. 4.11 **a** The ρ-T curves and **b** T_c as a function of RRR defined as $\rho(300\,\text{K})/\rho(T_c$ or $1.85\,\text{K})$ of Y_2O_2Bi samples. Samples for $1.1 \leq x \leq 1.6$ and $1.7 \leq x \leq 2.0$ are located in blue and green circles, respectively. **c** Sample dependence of carrier density derived from hall resistance versus magnetic field at 100 K

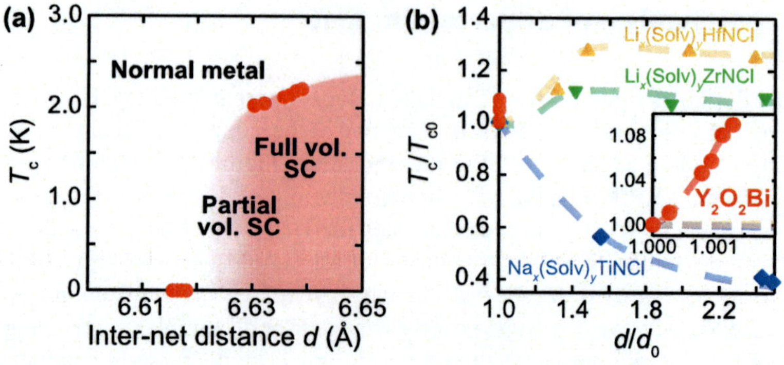

Fig. 4.12 **a** Phase diagram with T_c as a function of inter-net distance $d = c/2$. Partial and full volume superconductivity was defined from the shielding volume fraction (Fig. 4.3b). **b** The evolution of T_c as a function of normalized Bi^{2-} inter-net and MN inter-layer distance for Y_2O_2Bi and $A_x(Solv)_yMNCl$ compounds, respectively. T_{c0} and d_0 were defined from those of $x = 1.5$ in this study, and those of Li_xHfNCl, Li_xZrNCl, and Na_xTiNCl [41], respectively

tent,[2] which is consistent with the previous result [6] and electronic configuration of Bi^{2-}. Since the concentration of incorporated O has to be within several percent (~10^{20} cm^{-3}) considering the small c-axis expansion in comparison with the atomic intercalation into layered compounds [39], this result indicates the negligible role of the incorporated O as the carrier dopant.

An alternative scenario to induce superconductivity in Y_2O_2Bi is the enhanced two-dimensionality of the Bi^{2-} square net as a result of the abruptly expanded c-axis, because the expansion is concomitant with the emergence of the superconductivity for $1.5 \leq x \leq 2.0$. Figure 4.12a shows the T_c as a function of the inter-net distance d ($d = c/2$). No superconducting signatures were observed for $d \leq 6.62$ Å except for the slight resistance-drop at about 1.9 K in case of $d \approx 6.62$ Å ($x = 1.4$ in Fig. 4.6b). The partial volume superconductivity emerged at $d = 6.63$ Å with $T_c = 2.02$ K, evidenced by a small shielding volume fraction (Fig. 4.5b) and broad resistance-drop ($x = 1.5$ in Fig. 4.6b). Then, the full volume superconductivity emerged for $d > 6.63$ Å with a large shielding volume fraction (Fig. 4.5b) and sharp resistance-drop ($1.6 \leq x \leq 2.0$ in Fig. 4.6b). Here, T_c was approximately proportional to d. This result suggests that the superconductivity was induced by reducing inter-net coupling, i.e., enhancing two-dimensionality of the Bi^{2-} square net. The emergence of superconductivity via enhanced two-dimensionality of the conduction layer was scarcely seen in other superconductors, indicating that this concept would be a novel route to induce superconductivity in layered compounds.[3]

[2] Their very low Hall resistance with the significant scattering resulted in the large error bars and probably the overestimation of the carrier density.

[3] In the case of ionic and molecular intercalation into van der Waals gap as reported in graphite, chalcogenides, nitride halides, and so on, carrier doping or intercalated components were essential to induce the superconductivity [21], rather than increased inter-layer distance.

A similar tendency of the enhanced T_c without further carrier doping was observed in $A_x(Solv)_yMNCl$ (*A*: alkali metals; Solv: organic molecules; *M*: Ti, Zr, Hf) with increasing inter-layer distance *d* between double honeycomb *M*N layers by intercalating organic molecules into the van del Waals gap [22, 40, 41]. Figure 4.12b shows the relative change in T_c as a function of normalized Bi^{2-} inter-net and *M*N inter-layer distance for $Y_2O_2Bi$and $A_x(Solv)_yMNCl$, respectively. The T_c of Y_2O_2Bi increased much faster as a function of the normalized distance than those of $A_x(Solv)_yMNCl$ without saturation (inset of Fig. 4.12b), implying a peculiar feature of the superconductivity of monatomic Bi^{2-} square net.

4.6 Possible Homologous Series $Y_nO_nF_{n-2}Bi$ with Higher T_c

Observed T_c in polycrystalline Y_2O_2Bi was around 2 K, which was slightly lower than the onset T_c in Y_2O_2Bi epitaxial thin film ($T_c \approx 4$ K, Sect. 3.3.4). Enhanced T_c with increasing Bi^{2-} inter-net distance suggests the possible formation of homologous series $Y_nO_nF_{n-2}Bi$ (Fig. 4.13) because of the incidental F doping into the film (Sect. 3.3.2). The F doping was indispensable to retain the homologous framework for $n \geq 3$ due to the charge neutrality. The same framework of the blocking layer

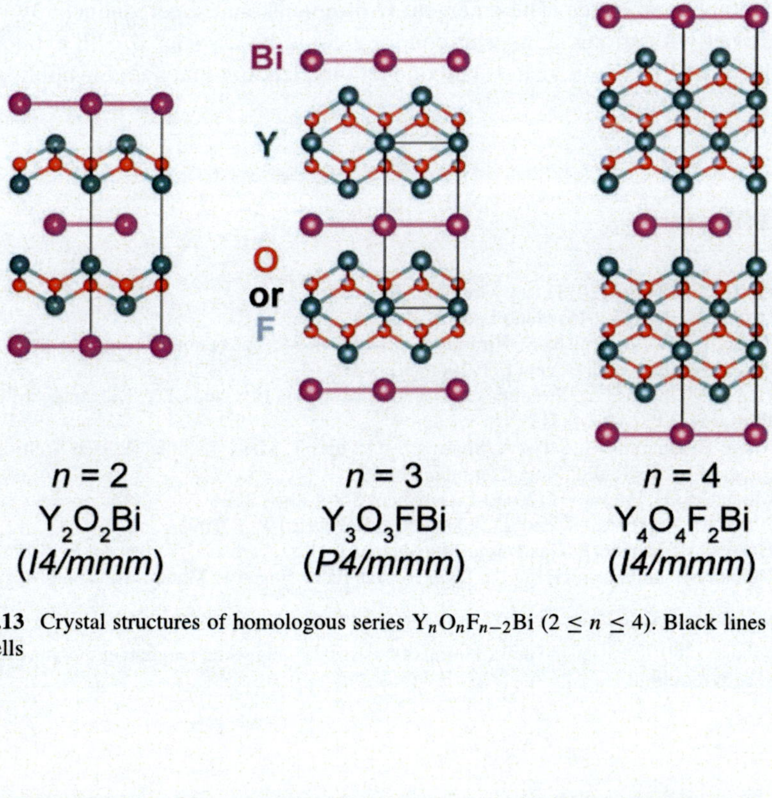

Fig. 4.13 Crystal structures of homologous series $Y_nO_nF_{n-2}Bi$ ($2 \leq n \leq 4$). Black lines denote unit cells

for $n = 3$ ($[Y_3O_3F]^{2+}$) was observed in $[Eu_3F_4]^{2+}$ layer of a structurally similar $Eu_3F_4Bi_2S_4$ [42].

In order to synthesize Y_3O_3FBi in the form of polycrystalline powder, a stoichiometric amount of mixed starting materials (Bi, Y, Y_2O_3, YF_3) was heated in an evacuated quartz tube at higher temperature than the usual case, 1100 and 1400°C,[4] which was applied to obtain other homologous series [43]. However, the products were just the mixture of Y_2O_2Bi, Y_2O_3, and YOF, indicating the difficulty to synthesize Y_3O_3FBi in bulk form. This is consistent with the very small superconducting volume in Y_2O_2Bi epitaxial thin film. An alternative technique such as high-pressure synthesis would be desired to achieve a homologous series $Y_nO_nF_{n-2}Bi$ phase.

4.7 Conclusion

I succeeded in inducing superconductivity at 2 K in Y_2O_2Bi which has been already concluded as non-superconducting in the previous reports. This is the fist report of superconducting Bi square net, which was evidenced by the BKT transition indicating two-dimensional superconductivity. The superconductivity emerged by increasing inter-net distance via O incorporation into hidden interstitial site without carrier doping, which was scarcely observed in other superconductors, suggesting an unconventional superconducting mechanism. These concepts would be a novel route to induce superconductivity in layered compounds, shedding light on hidden superconducting compounds. The emergent two-dimensional superconductivity of Bi^{2-} square net with strong spin–orbit coupling could be associated with exotic superconductivity like topological superconductivity, leading to a possible application in quantum computers.

References

1. Qi XL, Zhang SC (2011) Rev Mod Phys 83:1057
2. Luo W, Xiang H (2015) Nano Lett 15(5):3230
3. Mizoguchi H, Matsuishi S, Hirano M, Tachibana M, Takayama-Muromachi E, Kawaji H, Hosono H (2011) Phys Revi Lett 106(5):057002
4. Han F, Malliakas CD, Stoumpos CC, Sturza M, Claus H, Chung DY, Kanatzidis MG (2013) Phys Rev B 88(14):144511
5. Vinod K, Bharathi A, Satya A, Sharma S, Devidas T, Mani A, Sinha A, Deb S, Sridharan V, Sundar C (2014) Solid State Commun 192:60
6. Mizoguchi H, Hosono H (2011) J Am Chem Soc 133(8):2394
7. Rotter M, Tegel M, Johrendt D (2008) Phys Rev Lett 101:107006
8. Hosono H, Tanabe K, Takayama-Muromachi E, Kageyama H, Yamanaka S, Kumakura H, Nohara M, Hiramatsu H, Fujitsu S (2016) Sci Technolog Adv Mater

[4] Heating at 1400°C caused the deformation of quartz tube, suggesting higher temperature treatment would be impossible.

9. Nakashima Y, Yui H, Sasagawa T (2010) Phys C Supercond 470(20):1063
10. Takeshita N, Sasagawa T, Sugioka T, Tokura Y, Takagi H (2004) J Phys Soc Jpn 73(5):1123
11. Lin X, Straszheim WE, Bud'ko SL, Canfield PC (2013) J Alloys Compd 554:304
12. Kim SW, Lee K, Adroja D, Demmel F, Taylor J, Jung MH (2014) J Appl Phys 116(7):073901
13. Rosa P, Jesus C, Adriano C, Fisk Z, Pagliuso P (2015) J Phys Conf Ser (IOP Publishing):012063
14. Retzlaff R, Buckow A, Komissinskiy P, Ray S, Schmidt S, Mühlig H, Schmidl F, Seidel P, Kurian J, Alff L (2015) Phys Rev B 91:104519
15. Park J, Lee G, Wolff-Fabris F, Koh Y, Eom M, Kim Y, Farhan M, Jo Y, Kim C, Shim J et al (2011) Phys Rev Lett 107(12):126402
16. Wang K, Graf D, Lei H, Tozer S, Petrovic C et al (2011) Phys Rev B 84(22):220401
17. He J, Wang D, Chen G (2012) Appl Phys Lett 100(11):112405
18. Wang K, Wang L, Petrovic C (2012) Appl Phys Lett 100(11):112111
19. Izumi F, Momma K (2007) Applied crystallography XX, Solid State Phenomena, vol 130. Trans Tech Publications, pp 15–20. https://doi.org/10.4028/www.scientific.net/SSP.130.15
20. Kohama Y, Tojo T, Kawaji H, Atake T, Matsuishi S, Hosono H (2006) Chem Phys Lett 421(4):558
21. Klemm RA (2012) Layered superconductors, vol. 1. Oxford University Press
22. Yamanaka S, Hotehama KI, Kawaji H (1998) Nature 392(6676):580
23. Hor Y, Williams A, Checkelsky J, Roushan P, Seo J, Xu Q, Zandbergen H, Yazdani A, Ong N, Cava R (2010) Phys Rev Lett 104(5):057001
24. Sasaki S, Kriener M, Segawa K, Yada K, Tanaka Y, Sato M, Ando Y (2011) Phys Rev Lett 107(21):217001
25. Berezinskii V (1971) Soviet J Exp Theor Phys 32:493
26. Kosterlitz JM, Thouless DJ (1973) J Phys C Solid State Phys 6(7):1181
27. Halperin B, Nelson DR (1979) J Low Temp Phys 36(5–6):599
28. Yeh NC, Tsuei CC (1989) Phys Rev B 39:9708
29. Li Q, Hücker M, Gu GD, Tsvelik AM, Tranquada JM (2007) Phys Rev Lett 99:067001
30. Schneider R, Zaitsev A, Fuchs D et al (2014) J Phys Condens Matter 26(45):455701
31. Lin Z, Mei C, Wei L, Sun Z, Wu S, Huang H, Zhang S, Liu C, Feng Y, Tian H et al (2015) Sci Rep 5
32. Collan HK, Krusius M, Pickett GR (1969) Phys Rev Lett 23:11
33. Martin DL (1961) Proc R Soc London A Math Phys Eng Sci 263(1314):378
34. McMillan WL (1968) Phys Rev 167:331
35. Buslaev YA, Kravchenko E, Pachomov V, Skorikov V, Semin G (1969) Chem Phy Lett 3(6):455
36. Nanney CA, Garno JP (1972) Phys Rev Lett 28:1169
37. Robinson HG, Dehmelt HG, Gordy W (1953) Phys Rev 89:1305
38. Volovik G (1993) ZhETF Pisma Redaktsiiu 58:457
39. Morosan E, Zandbergen H, Dennis B, Bos J, Onose Y, Klimczuk T, Ramirez A, Ong N, Cava R (2006) Nat Phys 2(8):544
40. Takano T, Kishiume T, Taguchi Y, Iwasa Y (2008) Phys Rev Lett 100:247005
41. Kasahara Y, Kuroki K, Yamanaka S, Taguchi Y (2015) Phys C Supercond Appl 514:354
42. Zhai HF, Zhang P, Wu SQ, He CY, Tang ZT, Jiang H, Sun YL, Bao JK, Nowik I, Felner I et al (2014) J Am Chem Soc 136(43):15386
43. Ogino H, Sato S, Kishio K, Shimoyama JI, Tohei T, Ikuhara Y (2010) Appl Phys Lett 97(7):2506

Chapter 5
Unusual Superconductivity in Tb_2O_2Bi

Check for updates

5.1 Introduction

The superconductivity in Y_2O_2Bi was induced by the enhancement of two-dimensionality of Bi^{2-} square net via O incorporation, as described in the previous chapter. The next interesting question was the emergence of similar superconductivity in other R_2O_2Bi compounds. The $ThCr_2Si_2$-type Tb_2O_2Bi has been already synthesized, but the physical properties are not investigated [1]. Tb_2O_2Bi has a larger c-axis length than Y_2O_2Bi due to the larger ionic radii of Tb^{3+}, possibly connecting to the higher T_c. Furthermore, novel phenomena would be induced by the interaction between the superconductivity of Bi^{2-} square net and the magnetic moment of $[Tb_2O_2]^{2+}$.

In this chapter, polycrystalline Tb_2O_2Bi samples were successfully synthesized and their detailed physical properties were investigated for the first time. Pristine Tb_2O_2Bi showed a new electronic phase considered as CDW transition not observed in other R_2O_2Bi, in addition to the AF ordering of Tb^{3+}. The O incorporation was achieved by mixing CaO during synthesis, connecting to the emergence of superconductivity. Rich electronic phases suggest the unconventional superconductivity of Bi^{2-} square net in Tb_2O_2Bi.

5.2 Properties of Pristine Tb_2O_2Bi

5.2.1 Experimental

Polycrystalline Tb_2O_2Bi samples were synthesized by a similar procedure to that of Y_2O_2Bi. The starting materials were Bi (3N), Tb (3N), and Tb_4O_7 (3N) powders. The Tb_4O_7 powder was heated at 1000 °C for 1 day before use in order to eliminate absorbed moisture. A nominal amount of $Y_2O_{2-0.1n}Bi_{1+0.1n}$ (n: integer, $0 \leq n \leq 6$)

R. Sei, *Two-Dimensional Superconductivity in Rare Earth Oxybismuthides with Unusual Valent Bismuth Square Net*, Springer Theses,
https://doi.org/10.1007/978-981-99-7313-2_5

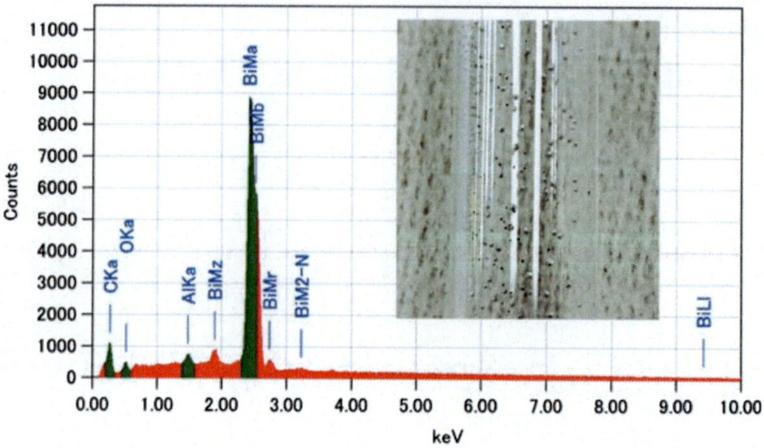

Fig. 5.1 Picture of the quartz tube after synthesis. Evaporated excess amount of Bi stuck to the inner wall of the tube, evidenced by EDX measurement. The signal of Al was originating from sample holder of EDX measurement

mixed powders was pressed into pellets by ~ 10 MPa and covered with Ta foil. The samples were denoted by using a nominal amount of O, i.e., x, instead of n. This operation was performed in a N_2-filled glovebox. The excess amount of Bi evaporated during the heating (Fig. 5.1), so the nominal Bi amount was also systematically varied. The pellets were heated in evacuated quartz tubes at 500 °C for 7.5 h, followed by heating at 1000 °C for 20 h. The products were ground, pressed into pellets by ~ 30 MPa, covered with Ta foil in the glovebox, and heated in evacuated quartz tubes at 1000 °C for 10 h. Crystal structures were evaluated by powder XRD using Cu Kα radiation. Rietveld refinements were performed by RIETAN-FP [2]. Magnetic properties were measured by SQUID magnetometer. Electrical transport properties were evaluated by the standard four-probe method.

5.2.2 *Crystal Structure*

Figure 5.2a shows a typical powder XRD pattern of Tb_2O_2Bi sample for $x = 1.6$. Most of diffraction peaks were completely fitted by simulation as $ThCr_2Si_2$ -type Tb_2O_2Bi phase, indicating its successful synthesis. Refined structural parameters and phase purity are summarized in Table 5.1 and Fig. 5.2b. Several non-superconducting minor phases appeared depending on the prescribed amount of O: TbBi for reductive condition ($x = 1.4, 1.5$) and Tb_2O_3 for oxidative condition ($1.7 \leq x \leq 2.0$), similar to Y_2O_2Bi. However, the lattice constants showed almost constant values regardless of x (inset of Fig. 5.2 and Table 5.1), which are quite similar values to those in the previous report ($a = 3.8962$, $c = 13.317$ Å) [1] indicating their ideal composition.

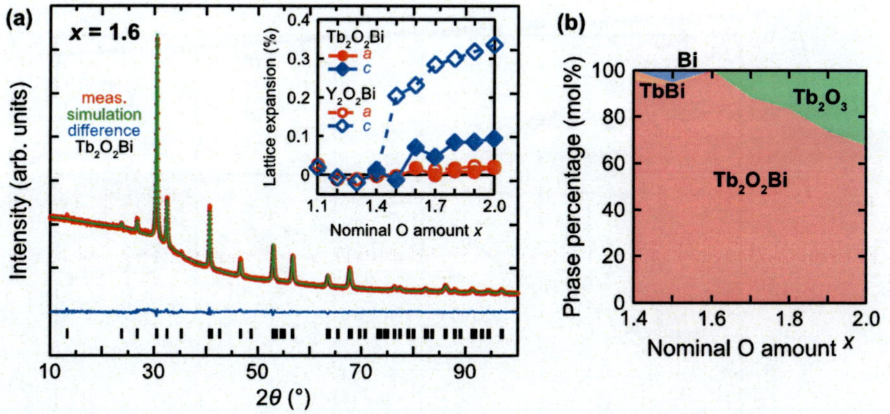

Fig. 5.2 **a** Typical powder XRD pattern of Tb_2O_2Bi for $x = 1.6$ with Rietveld refinement. Inset shows the rate of lattice constant change in Tb_2O_2Bi and Y_2O_2Bi samples. **b** Phase diagram for Tb_2O_2Bi samples

Table 5.1 Summary of structural data in Tb_2O_2Bi samples

x	1.4	1.5	1.6	1.7	1.8	1.9	2.0
Space group	*I4/mmm*	*I4/mmm*	*I4/mmm*	*I4/mmm*	*I4/mmm*	*I4/mmm*	*I4/mmm*
a (Å)	3.89932(14)	3.8988(3)	3.8997(2)	3.8990(5)	3.8994(3)	3.8993(3)	3.8999(3)
c (Å)	13.3183(4)	13.3150(11)	13.3261(7)	13.3228(18)	13.3278(11)	13.3280(10)	13.3291(9)
V (Å^3)	202.501(12)	202.39(3)	202.66(2)	202.54(5)	202.65(3)	202.65(3)	202.72(2)
Tb z	0.32969(18)	0.3318(3)	0.3290(2)	0.3285(3)	0.3291(3)	0.3300(3)	0.3290(4)
Purity (mol%)	95.9	93.4	> 99	88.4	83.4	73.8	67.8
R_{wp}	1.401	1.308	1.465	1.310	1.286	1.199	1.191
R_e	1.029	0.997	0.981	0.982	0.962	0.952	0.944
S	1.3620	1.3121	1.4939	1.3344	1.3366	1.2586	1.2609

Wyckoff positions: Bi on the $2a$ site (0, 0, 0), Tb on the $4e$ site (0, 0, z), and O on the $4d$ site (0, 0.5, 0.25)

This result indicates the lack of O incorporation in Tb_2O_2Bi synthesized even in oxidative condition, contrary to the case of Y_2O_2Bi.[1]

5.2.3 Physical Properties

Here, I showed physical properties of the highest purity sample $x = 1.6$, because of no qualitative change among x variation. Novel electronic phases appeared although

[1] Tb_2O_2Bi sample synthesized from Bi rich starting materials (nominal composition: $Tb_2O_2Bi_{1.5}$) showed similar lattice constants ($a = 3.8999(3)$, $c = 13.3296(12)$ Å), supporting that the absence of c-axis expansion was originating from the lack of not Bi but O incorporation (see also Sect. 5.3.3).

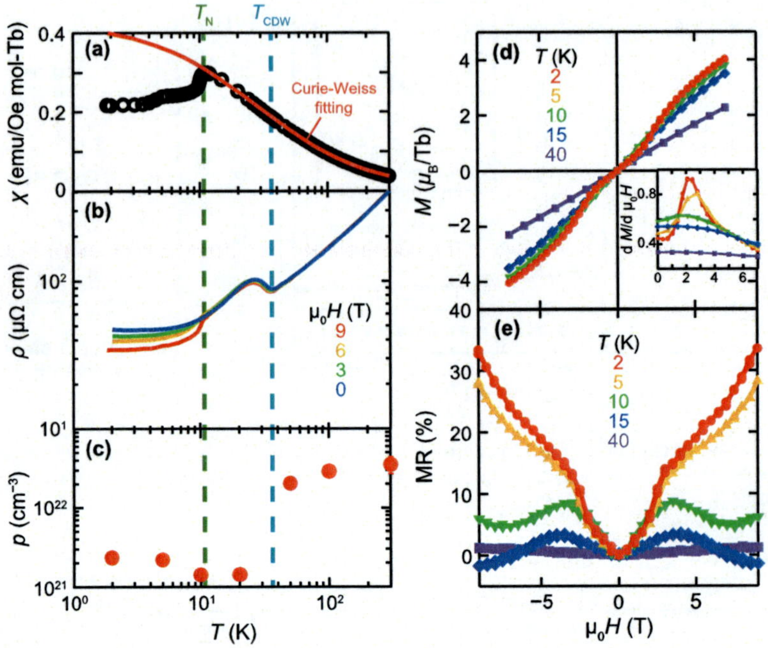

Fig. 5.3 Physical properties of Tb_2O_2Bi for $x = 1.6$. Temperature dependence of **a** magnetic susceptibility, **b** electrical resistivity, and **c** carrier density. Broken lines denote Néel temperature (green) and transition temperature of CDW (blue). **d** Magnetization curve with first derivative (inset) and **e** MR at various temperatures

superconductivity was not observed down to 1.8 K probably due to the lack of c-axis expansion. Figure 5.3a shows χ-T curve in the FC process at 100 Oe. Sudden χ drop at 11.1 K indicates AF transition of Tb_2O_2 layer, similarly observed in other magnetic R_2O_2Bi [3].[2] Above Néel temperature ($T > T_N$), χ-T curve was completely fitted by the Curie–Weiss law $\chi = \chi_0 + C/(T - \theta)$, and the obtained parameters are summarized in Table 5.2. The effective Bohr magneton values P_{eff} can be calculated from the relationships $C = Nm^2/3k_B$ and $\mu_B = 9.274 \times 10^{-21}$ erg Oe^{-1}, where N is the number of magnetic atoms, m is the magnetic moment, k_B is the Boltzmann constant, and μ_B is the Bohr magneton [5]. Obtained $P_{eff} = 9.86\mu_B$ coincided with theoretical values $9.72\mu_B$ for Tb^{3+},[3] confirming trivalent state of Tb cations similar to other R_2O_2Bi [3]. Figure 5.3d shows the magnetization curve. An almost linear behavior of magnetization was observed owing to the paramagnetic nature of local spin at $T > T_N$. A sudden increase in magnetization was observed around ±2T at $T < T_N$ evidenced by the first derivative curves (inset of Fig. 5.3d). These kinks

[2] A small χ drop at 5.1 K was originating from impurity $Tb_{11}O_{20}$ [4].

[3] Theoretical values of P_{eff} in R^{3+} cations are 3.58 (Pr^{3+}), 7.94 (Gd^{3+}), 9.72 (Tb^{3+}), 10.64 (Dy^{3+}), and $9.58\mu_B$ (Er^{3+}), respectively [5].

Table 5.2 Magnetic properties for R_2O_2Bi

R_2O_2Bi	T_N (K)	χ_0 (emu mol-R^{-1})	θ (K)	C (emu mol-R^{-1})	P_{eff} (μ_B)
Pr_2O_2Bi [3]	15.0	0.0008	−25.1	1.71	3.70
Gd_2O_2Bi [3]	10.1	0.0011	−25.0	7.80	7.90
Tb_2O_2Bi	11.1	0.0008	−28.5	12.15	9.86
Dy_2O_2Bi	10.8	0.0005	−25.3	14.75	10.86
Er_2O_2Bi [3]	3.0	0.0021	−19.4	12.32	9.90

See Appendix B

were attributed to the metamagnetic transition which is well observed in rare earth compounds [6].

Figure 5.3b shows ρ-T curves at various magnetic fields. A ρ drop at 11 K was originating from the AF ordering in $[Tb_2O_2]^{2+}$ layers similar to that observed in other AF Bi square net compounds [7–9], resulting in giant MR up to ~ 34% at 2 K and 9 T (Fig. 5.3e). In addition, a broad hump was observed around 30 K as was never seen in R_2O_2Bi compounds [3] without any indication in χ-T curve, suggesting CDW transition of Bi^{2-} square net layers which is frequently observed in low-dimensional layered compounds such as transition metal chalcogenides [10–14]. It is considered that Bi^{2-} anions formed dimers at $T < T_{CDW}$ in Tb_2O_2Bi, leading to a transition from square net to ladder-type structure with new periodicity (Fig. 5.4), similar to the results in analogous R_2O_2Sb [15–19].[4] The generation of new structural periodicity

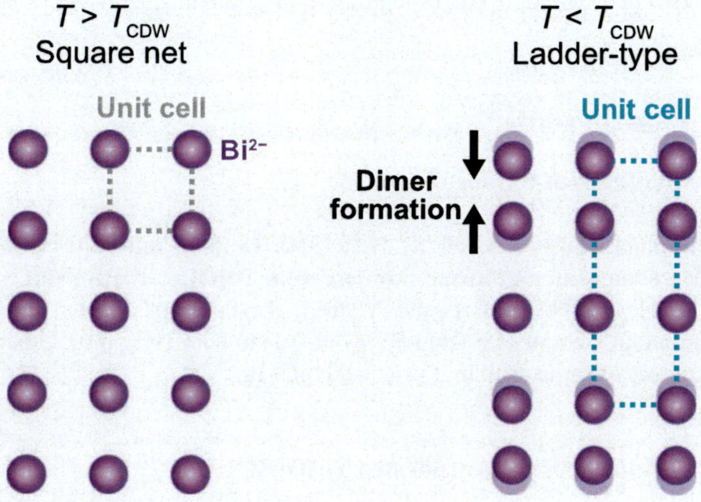

Fig. 5.4 A schematic image of CDW transition from the top view of Bi^{2-} layer in Tb_2O_2Bi

[4] XRD measurement at 4 K could not detect the direct evidence of structural transition, indicating quite small distortion of Bi^{2-} square net.

opened the gap in band dispersion of Tb_2O_2Bi, resulting in the dramatic decrease of hole carrier density at $T < T_{CDW}$ derived from Hall measurement (Fig. 5.3c).

Figure 5.3e shows MR at various temperatures. Generally, the MR behavior was determined by the shape of the Fermi surface [20], so the change of MR trend at $T < T_{CDW}$ demonstrated the Fermi surface reconstruction by CDW transition. One might consider that this anomaly was originating from the interaction between Kondo and crystalline-electric-field effects observed in $CeNiBi_2$ and $CePd_{1-x}Bi_2$ [21, 22]. However, the Curie–Weiss-type χ-T curve eliminated the possibility of this scenario.

It is noted that both AF and CDW phases usually compete with the emergence of superconductivity. Furthermore, the emergence of CDW transition contradicted the theoretical prediction which indicates the absence of CDW transition in R_2O_2Bi compounds due to the strong spin–orbit coupling and chemical pressure [19].

5.3 Structural Control of Tb_2O_2Bi

Tb_2O_2Bi did not show c-axis expansion by the same synthesis procedure as Y_2O_2Bi, resulting in the absence of superconductivity. Instead, unexpected CDW transition of Bi^{2-} square net emerged with AF ordering of $[Tb_2O_2]^{2+}$ layer. The CDW transition can be suppressed by applying pressure in two-dimensional compounds such as transition metal chalcogenides because their Fermi surface approaches three-dimensional nature, occasionally connecting to the emergence of superconductivity [13, 14]. In this study, I controlled the Bi^{2-} inter-net distance in Tb_2O_2Bi by F substitution for O site and O incorporation, corresponding to the uniaxial chemical pressure.

5.3.1 Experimental

5.3.1.1 Synthesis of F Doped Tb_2O_2Bi

The starting materials were Bi (3N), BiF_3 (3N), Tb (3N), and Tb_4O_7 (3N) powders. A nominal composition of mixed powders were $Tb_2O_{1.6-y}F_yBi_{1.4}$ ($0.1 \leq y \leq 0.6$) (Table 5.3) because $Tb_2O_{1.6}Bi_{1.4}$ showed the highest purity. I denoted the samples by using a nominal amount of F from 5% ($y = 0.1$) to 30% ($y = 0.6$). Other synthesis procedure was as same as that of pristine Tb_2O_2Bi.

5.3.1.2 Synthesis of O Incorporated Tb_2O_2Bi

In order to achieve O incorporation into Tb_2O_2Bi, further an oxidative method was developed by mixing CaO during synthesis. The CaO powders were prepared by heating $CaCO_3$ at 1000 °C for 10h in the air, whose single phase was verified by XRD measurement. Other synthesis procedure was as same as that of pristine Tb_2O_2Bi and

Table 5.3 Summary of structural data in F doped Tb_2O_2Bi samples

F content (%)	5	10	15	20	25	30
Nominal composition	$Tb_2O_{1.5}F_{0.1}Bi_{1.4}$	$Tb_2O_{1.4}F_{0.2}Bi_{1.4}$	$Tb_2O_{1.3}F_{0.3}Bi_{1.4}$	$Tb_2O_{1.2}F_{0.4}Bi_{1.4}$	$Tb_2O_{1.1}F_{0.5}Bi_{1.4}$	$Tb_2O_{1.0}F_{0.6}Bi_{1.4}$
a (Å)	3.89959(18)	3.90043(12)	3.90379(13)	3.90550(13)	3.90608(14)	3.90569(12)
c (Å)	13.3117(6)	13.2894(4)	13.1777(4)	13.1336(4)	13.1208(4)	13.1221(4)
V ($Å^3$)	202.429(16)	202.176(11)	200.823(12)	200.326(11)	200.190(12)	200.170(10)
Tb z	0.3306(3)	0.3330(2)	0.33479(19)	0.33574(16)	0.3366(2)	0.3368(2)
Purity (mol%)	85.1	80.6	84.5	84.1	69.6	60.9
R_{wp}	1.331	1.328	1.404	1.656	1.772	1.665
R_e	1.034	1.032	1.023	1.030	1.018	1.053
S	1.2833	1.2867	1.3729	1.6073	1.7400	1.5809

a nominal composition of starting materials are summarized in Table 5.4. I denoted the samples by using sample # in Table 5.4. At high temperature during synthesis, CaO served as an O supplier through self-reduction to Ca metal with strong reductant Tb. The generated O from CaO would be incorporated into Tb_2O_2Bi framework, and Ca residue evaporated because of higher vapor pressure than Bi resulting in non-doping into Tb_2O_2Bi framework. This was evidenced by SEM-EDX measurements of both quartz tube after synthesis and the sample pellet (Fig. 5.5).

5.3.2 *Crystal Structure*

Figure 5.6a shows the typical powder XRD pattern of F doped Tb_2O_2Bi sample for 20% of F content. Most of diffraction peaks were fitted by simulation as $ThCr_2Si_2$-type Tb_2O_2Bi phase with several non-superconducting impurity phases such as TbOF and Bi (Fig. 5.6b). Refined structural parameters are summarized in Table 5.3. The *a*-axis length showed almost constant values. On the other hand, the *c*-axis length monotonically decreased with increasing F content (inset of Fig. 5.6), indicating substitution of F for O site as observed in Y_2O_2Bi epitaxial thin film and structurally similar *R*FeAsO and *R*OBiS$_2$ [23, 24]. The decrease in *c*-axis length was almost saturated around 20% of F content with an increase of impurity phases, indicating the solubility limit of F was around 20%.

Figure 5.7a shows a typical powder XRD pattern of Tb_2O_2Bi sample synthesized with CaO (sample #3). Most of the diffraction peaks were fitted by simulation as $ThCr_2Si_2$-type Tb_2O_2Bi phase with Bi, and Tb_2O_3 impurity phases supported the preferential evaporation of Ca and strong oxidative condition, respectively, as explained above.[5] Refined structural parameters are summarized in Table 5.4. All Tb_2O_2Bi samples synthesized with CaO showed *c*-axis expansion with almost constant *a*-axis length compared with pristine Tb_2O_2Bi ($a = 3.8997$Å, $c = 13.3261$Å, for $x = 1.6$), indicating the achievement of O incorporation into Tb_2O_2Bi (inset of Fig. 5.7). These results indicated the successful chemical control of Tb_2O_2Bi structure by substitution of F for O site and O incorporation, corresponding to the application of positive and negative uniaxial chemical pressure.

5.3.3 *Physical Properties of F Doped Tb_2O_2Bi*

Magnetic properties of F doped Tb_2O_2Bi were qualitatively similar to those of pristine Tb_2O_2Bi, that is, the AF ordering and metamagnetic transition at $T < T_N$ was observed in spite of a small variation of T_N as summarized in Table 5.5. Notable effect

[5] CaO appeared as an impurity when a large amount of CaO was mixed in starting materials such as #5 and #6. However, quantitative analysis by Rietveld refinement did not go well due to the low quality of XRD data.

Table 5.4 Summary of structural data in O incorporated Tb_2O_2Bi samples

Sample #	1	2	3	4	5	6
Nominal composition	$Tb_2O_{1.8}Bi_{1.4} + (CaO)_{0.2}$	$Tb_2O_{1.6}Bi_{1.6} + (CaO)_{0.2}$	$Tb_2O_{1.8}Bi_{1.4} + (CaO)_{0.4}$	$Tb_2O_{1.6}Bi_{1.4} + (CaO)_{0.4}$	$Tb_2O_{1.6}Bi_{1.4} + (CaO)_{0.6}$	$Tb_2O_{1.2}Bi_{2.0} + (CaO)_{2.0}$
a (Å)	3.8992(4)	3.8997(4)	3.8991(5)	3.8996(5)	3.8994(4)	3.8987(4)
c (Å)	13.3488(13)	13.3542(13)	13.3579(17)	13.3638(15)	13.3649(15)	13.376(2)
V (Å^3)	202.95(3)	203.09(3)	203.08(4)	203.22(4)	203.22(4)	203.31(5)
Tb z	0.3272(4)	0.3286(3)	0.3279(4)	0.3285(5)	0.3275(3)	0.3221(5)
Purity (mol%)	47.4	54.3	72.2	55.9	55.3	34.4
R_{wp}	1.253	1.241	1.297	1.191	1.281	2.03
R_e	0.827	0.997	0.814	0.853	0.833	0.872
S	1.5152	1.4952	1.5942	1.3968	1.5381	2.3086

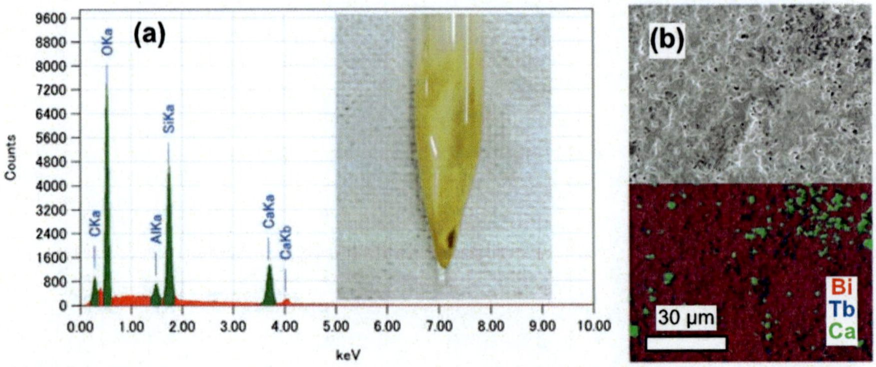

Fig. 5.5 **a** Picture and EDX spectrum of the quartz tube after synthesis of CaO mixed Tb_2O_2Bi (sample #5). **b** SEM image (top) and EDX mapping (bottom) of sample #5. Isolated Ca lumps evidenced non-doping into Tb_2O_2Bi framework

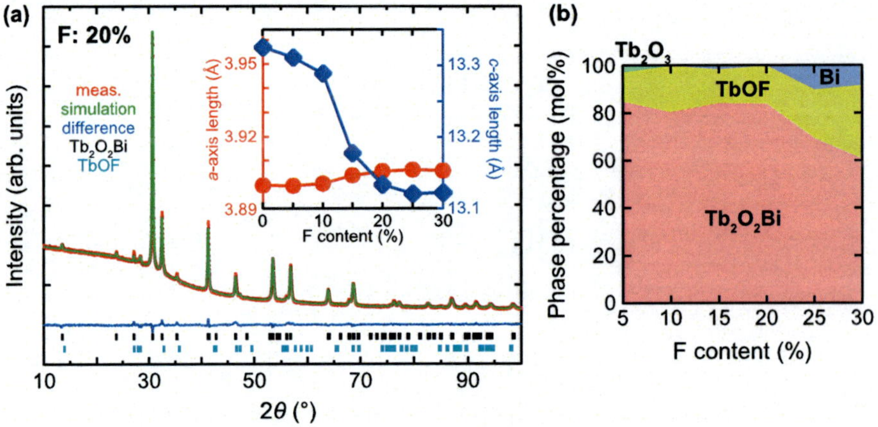

Fig. 5.6 **a** Typical powder XRD pattern of F doped Tb_2O_2Bi (F: 20%) with Rietveld refinement. Inset shows the change of lattice constant. The data for 0% of F content was pristine Tb_2O_2Bi for $x = 1.6$. **b** Phase diagram for F doped Tb_2O_2Bi samples

of positive uniaxial chemical pressure was observed in electrical transport properties (Fig. 5.8). The T_{CDW} shifted to lower temperature with increasing F content, and was almost suppressed for F content of 20% although superconductivity was not observed (Fig. 5.8a). This was also evidenced by the temperature dependence of carrier density: the carrier density dropped due to the gap opening in synchronous with T_{CDW} and was finally constant without suppression for F content of 20% (Fig. 5.8). Figure 5.8c–e shows MR at various temperatures for F doped Tb_2O_2Bi samples. The change of MR behavior suggests the Fermi surface evolution due to the effect of uniaxial chemical pressure.

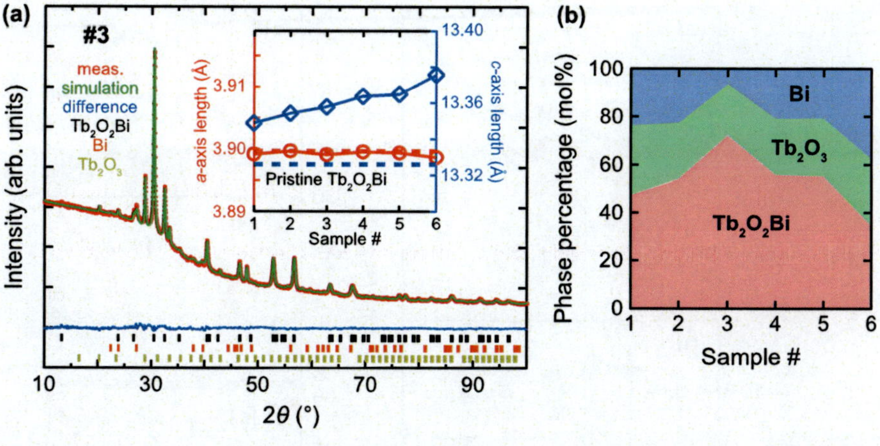

Fig. 5.7 **a** Typical powder XRD pattern of O incorporated Tb_2O_2Bi sample #3 with Rietveld refinement. Inset shows the change of lattice constant. Broken lines denote the data of pristine Tb_2O_2Bi for $x = 1.6$. **b** Phase diagram for O incorporated Tb_2O_2Bi samples

Table 5.5 Magnetic properties of F doped Tb_2O_2Bi

F content (%)	T_N (K)	χ_0 (emu mol-R^{-1})	θ (K)	C (emu mol-R^{-1})	P_{eff} (μ_B)
0	11.1	0.0008	−28.5	12.15	9.86
10	9.9	0.0028	−22.4	11.47	9.58
20	10.3	0.0021	−25.1	11.80	9.71

Carrier doping is also a powerful way to suppress the CDW transition [11]. Electron carriers from F substitution should have influenced the electrical transport properties of Tb_2O_2Bi, as observed in the polarity change of carriers at several points (open symbols in Fig. 5.8b). In order to consider the influence of carrier doping on the CDW transition, c-axis shrunk Tb_2O_2Bi was synthesized as a referential experiment by introducing Bi vacancy. The nominal composition of the sample was $Tb_2O_{2.0}Bi_{0.8}$ and obtained lattice constants were $a = 3.8973$ Å and $c = 13.1925$ Å, which were quite similar to those of 15% F doped Tb_2O_2Bi (Table 5.3). Figure 5.9 shows the comparison of ρ-T curves and MR between 15% F doped Tb_2O_2Bi and $Tb_2O_{2.0}Bi_{0.8}$. The T_{CDW} and MR behavior of $Tb_2O_{2.0}Bi_{0.8}$ were quite similar to those of 15% F doped Tb_2O_2Bi with decreased T_{CDW}, although there would be a small contribution of carrier doping and scattering at Bi vacancy to conductivity such as smaller RRR for $Tb_2O_{2.0}Bi_{0.8}$. This result indicates that the decrease of c-axis length was the driving force to suppress the CDW transition in Tb_2O_2Bi.

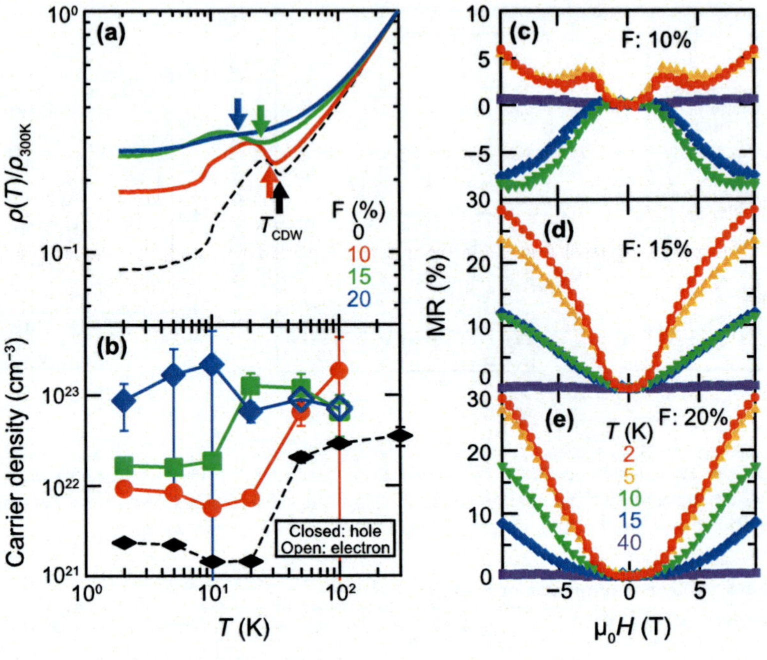

Fig. 5.8 Temperature dependence of **a** normalized electrical resistivity and **b** carrier density for F doped Tb_2O_2Bi. Arrows in **a** denote T_{CDW}. Closed (open) symbols in **b** denote hole (electron) carrier. **c–e** MR in various F doped Tb_2O_2Bi

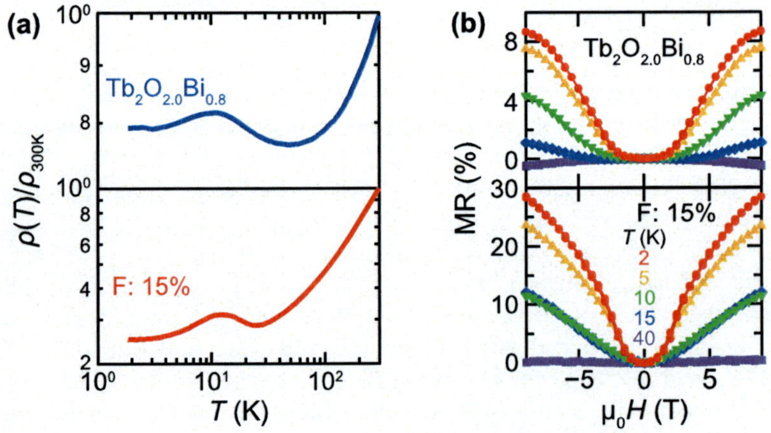

Fig. 5.9 **a** The ρ-T curves and **b** MR of $Tb_2O_{2.0}Bi_{0.8}$ (top) and 15% F doped Tb_2O_2Bi (bottom), respectively

5.3.4 Physical Properties of O Incorporated Tb_2O_2Bi

The effect of negative uniaxial chemical pressure for O incorporated Tb_2O_2Bi appeared remarkably in electrical transport properties as in the F doped Tb_2O_2Bi.[6] Figure 5.10a shows normalized ρ-T curves for O incorporated Tb_2O_2Bi. The hump originating from the CDW transition was gradually suppressed with a gradual decrease of T_{CDW} by enlarging c-axis length, leading to the concomitant emergence of superconductivity at about 2 K (right bottom inset of Fig. 5.10a). A large diamagnetic signal corresponding to the Meissner effect was observed at similar T_c in the χ-T curve (Fig. 5.10b), confirming bulk superconductivity of O incorporated Tb_2O_2Bi. The inset of Fig. 5.10b indicates type-II superconductivity with $H_{c1} \approx 5$ Oe and obtained H_{c2} was 353 Oe for sample #6 (left top inset of Fig. 5.10). The T_c was almost positively correlated with the c-axis length, indicating that the superconductivity in Tb_2O_2Bi was induced by enhanced two-dimensionality of Bi^{2-} square net although the T_c showed similar values to those of Y_2O_2Bi in spite of the further enlarged Bi^{2-} inter-net distance. In addition, smaller H_{c1} and H_{c2} than those of Y_2O_2Bi probably due to the large magnetic moment of Tb^{3+} (Tables 5.6 and 5.7).

The possibility of superconducting impurities can be completely eliminated by the following discussion. First, superconducting Bi was not responsible because of the different T_c as explained in Sect. 3.3.4. Second, suspects were Ca–Bi alloy compounds. There were three superconducting Ca–Bi alloys; $CaBi_3$ [25], $CaBi_2$ [26], and $Ca_{11}Bi_{10-x}$ [27]. The T_c of $CaBi_3$ is 1.7 K, lower than those in Tb_2O_2Bi samples. $CaBi_2$ shows similar $T_c = 1.95$ K, but it is a type-I superconductor. $Ca_{11}Bi_{10-x}$

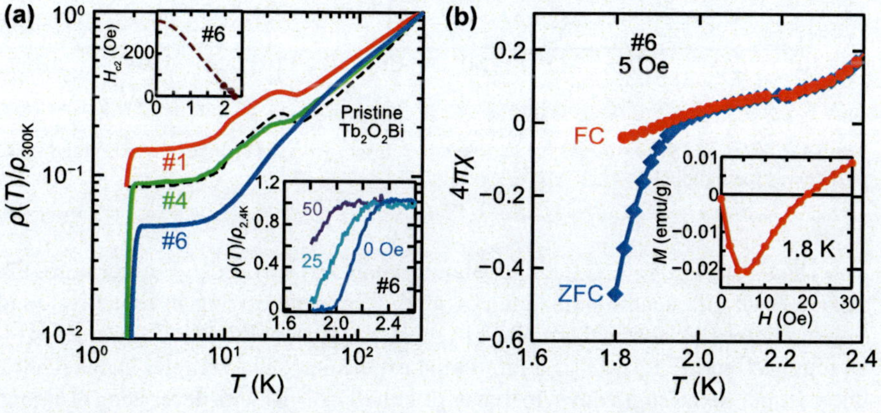

Fig. 5.10 **a** The ρ-T curves of O incorporated Tb_2O_2Bi. Right bottom inset shows magnified image of the ρ-T curve at low temperature and left top inset shows H_{c2} fitted by Ginzburg–Landau theory [28] for sample #6. **b** The χ-T curve for sample #6. Inset shows magnetization curve at 1.8 K

[6] AF signal overlapped with paramagnetism of Tb_2O_3 impurity (Fig. 5.7b), possibly hampered the correct evaluation.

Table 5.6 Physical properties of O incorporated Tb_2O_2Bi

Sample #	c (Å)	T_c (K)	T_N (K)	T_{CDW} (K)
Pristine ($x = 1.6$)	13.3261(7)	0	11.1	33.5
1	13.3488(13)	1.92	10.2	30.1
2	13.3542(13)	1.99	10.3	30.0
3	13,3579(17)	1.90	10.3	30.0
4	13.3638(15)	2.02	9.8	25.5
5	13.3649(15)	1.96	10.0	27.0
6	13.376(2)	2.29	–	–

Table 5.7 Nominal composition of the samples in referential experiment

Sample #	R1	R2	R3	R4
Composition	$Tb_2O_{1.6}Bi_{1.6}$ + $(CaO)_{0.2}$	$Tb_2O_{1.5}Bi_{1.6}$ + $(CaO)_{0.2}$	$Tb_2O_{1.4}Bi_{1.3}$ + $(CaO)_{0.2}$	$Tb_2O_{1.2}Bi_{1.8}$ + $(CaO)_{0.2}$

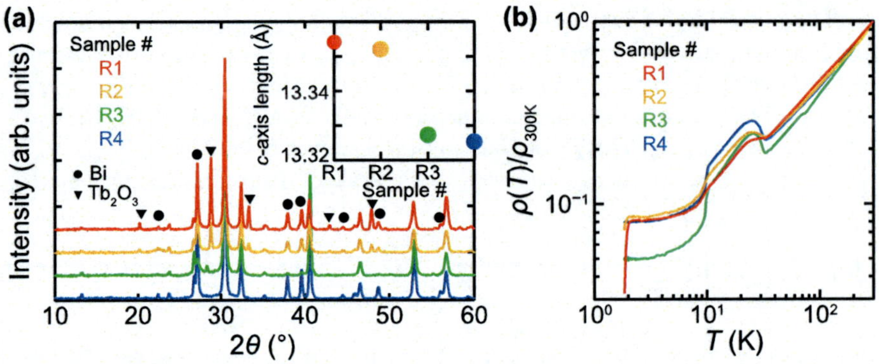

Fig. 5.11 **a** Powder XRD patterns for sample R1–4. Inset shows the change of c-axis length. **b** The ρ-T curves for sample R1–4

also shows similar $T_c = 2.2$ K, but shows small $H_{c2} = 172$Oe. Furthermore, CaO mixed Tb_2O_2Bi became non-superconducting by synthesizing in reductive condition as follows. A nominal amount of O was systematically varied (samples R1–4) as summarized in Fig. 5.11 as a referential experiment. The samples showed abrupt decrease of c-axis length down to that of pristine Tb_2O_2Bi with decreasing O amount, resulting in the suppression of superconductivity in samples R3 and R4 (Fig. 5.11). This result supports that c-axis expansion via O incorporation was essential to induce superconductivity in Tb_2O_2Bi.

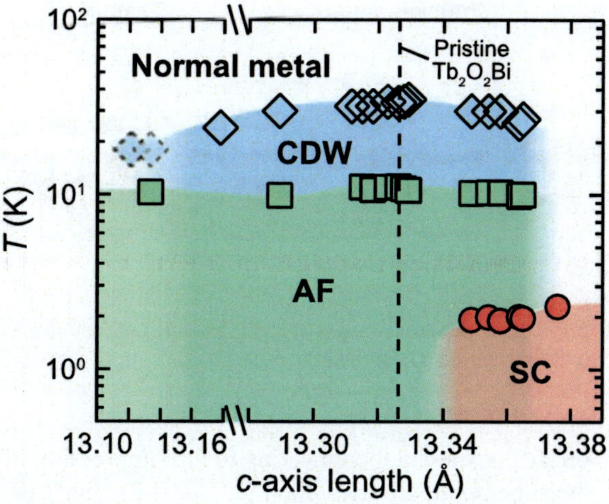

Fig. 5.12 Electronic phase diagram of Tb_2O_2Bi

5.3.5 *Complex Phase Diagram*

Figure 5.12 shows the electronic phase diagram of Tb_2O_2Bi as a function of *c*-axis length. The CDW transition was suppressed by decreasing *c*-axis length, similar to the pressurized transition metal chalcogenides [10, 12–14], although the superconductivity did not emerge. The inverse effect, that is the enlarging *c*-axis length, also suppressed the CDW transition, notably connecting to the emergence of superconductivity as observed in Y_2O_2Bi. Furthermore, AF and CDW phases coexisted with the superconductivity in the boundary region ($13.34 < c < 13.37$ Å), followed by the complete suppression of AF and CDW phases with the enhanced T_c for $c > 13.37$ Å. A similar complex electronic phase diagram was observed in cuprate [29, 30] and Fe-based [31, 32] high-temperature superconductors. These features suggest the unconventional superconductivity of Bi^{2-} square net in Tb_2O_2Bi.

Hereafter, I discuss the reason why the complex electronic phase diagram was observed in Tb_2O_2Bi. In the $AMnBi_2$ (*A*: Ca and Sr), the properties of Bi^- square net can be engineered by tuning the surrounding environment of Bi^- square net [33, 34]: a different atomic arrangement adjacent to Bi^- square net affects the degeneracy of the Dirac cone-like band dispersion in the single Bi square net, and different *A*-site elements, i.e., Sr or Ca, change the spin–orbit coupling effect resulting in the different modification of Fermi surface (Fig. 5.13). Furthermore, it was theoretically proposed that the CDW transition was suppressed in R_2O_2Bi due to the chemical pressure effect and strong spin–orbit coupling [19], contrary to my observation. Accordingly, it was expected that the CDW transition emerged due to reduced spin–orbit coupling of Bi^{2-} square net by the presence of Tb^{3+} for the same reason. It was possible that the CDW was suppressed by decreasing *c*-axis length because of the strong chemical pressure effect surpassing the reduced spin–orbit coupling effect. On the other hand, the expansion of *c*-axis length would eliminate the influence of Tb^{3+}

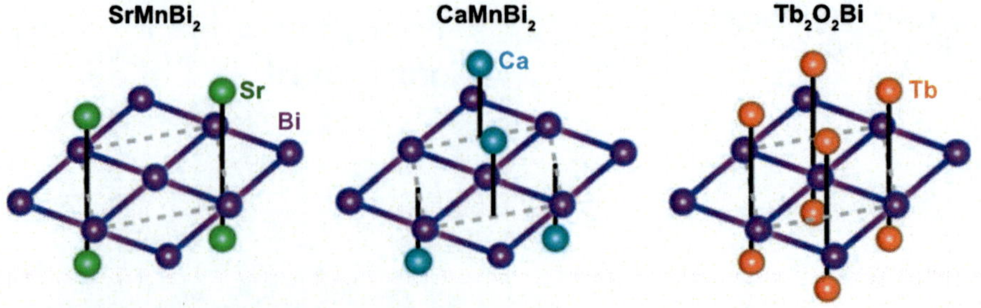

Fig. 5.13 Schematic images of local structure adjacent to Bi square net in $SrMnBi_2$ (left), $CaMnBi_2$ (center), and Tb_2O_2Bi (right)

on Bi^{2-} square net, connecting to the suppression of the CDW and the emergence of the superconductivity, which would be the intrinsic properties of a single Bi square net as the Dirac cone-like band dispersion.

The reason why only Tb^{3+} among the other R^{3+} had a different influence on the spin–orbit coupling of Bi^{2-} square net was unclear. One possible reason was the different spin arrangement in $[Tb_2O_2]^{2+}$ layer from the other $[R_2O_2]^{2+}$ layer. Both $CaMnBi_2$ and $SrMnBi_2$ show AF ordering of the MnBi layer, but inter-layer coupling is different between the ferromagnetic order in $CaMnBi_2$ and AF order in $SrMnBi_2$, resulting in the ρ-T kink at T_N only observed in $CaMnBi_2$ [7, 9]. The ρ drop at T_N was noticeable in Tb_2O_2Bi compared with Dy_2O_2Bi (Appendix B) in this study, suggesting a similar spin arrangement variation to $AMnBi_2$. This would be verified by the neutron diffraction measurement [7, 35].

5.4 Conclusion

I investigated the physical properties of Tb_2O_2Bi for the first time. Pristine Tb_2O_2Bi showed rich electronic phenomena such as the CDW transition of Bi^{2-} square net and the AF ordering of $[Tb_2O_2]^{2+}$ layer connecting to the giant MR, not observed in other R_2O_2Bi. However, the c-axis expansion via O incorporation was not observed just by varying a nominal amount of O in starting materials resulting in the absence of superconductivity. The electronic properties of Tb_2O_2Bi was controllable by applying uniaxial chemical pressure along c-axis. The c-axis shrink was obtained by substitution of F for O site, connecting to the suppression of the CDW transition despite the absence of superconductivity. On the other hand, the c-axis expansion via O incorporation was achieved under further oxidative synthesis condition by mixing CaO into starting materials. Superconductivity was observed with similar T_c trend to Y_2O_2Bi, associated with the suppression of the CDW transition. Furthermore, the superconductivity coexisted with the CDW and AF phases in the boundary region, suggesting unconventional superconductivity of Bi^{2-} square net in Tb_2O_2Bi.

References

1. Nuss J, Jansen M (2012) Zeitschrift für anorganische und allgemeine Chemie 638(3–4):611
2. Izumi F, Momma K (2007) Applied crystallography XX, solid state phenomena, vol 130. Trans Tech Publications, pp 15–20. https://doi.org/10.4028/www.scientific.net/SSP.130.15
3. Mizoguchi H, Hosono H (2011) J Am Chem Soc 133(8):2394
4. Baran S, Duraj R, Hoser A, Penc B, Szytuła A (2013) Acta Physica Polonica A 123(1):98
5. Chikazumi S (2008) Phys Ferromagn Mater (Shokabo)
6. Gignoux D, Schmitt D (1991) J Magn Magn Mater 100(1–3):99
7. Guo Y, Princep A, Zhang X, Manuel P, Khalyavin D, Mazin I, Shi Y, Boothroyd A (2014) Phys Rev B 90(7):075120
8. Adriano C, Rosa P, Jesus C, Grant T, Fisk Z, Garcia D, Pagliuso P (2015) J Appl Phys 117(17):17C103
9. Ray SJ, Alff L (2016) Phys Status Solidi (B)
10. DiMasi E, Aronson M, Mansfield J, Foran B, Lee S (1995) Phys Rev B 52(20):14516
11. Morosan E, Zandbergen H, Dennis B, Bos J, Onose Y, Klimczuk T, Ramirez A, Ong N, Cava R (2006) Nat Phys 2(8):544
12. Ru N, Condron C, Margulis G, Shin K, Laverock J, Dugdale S, Toney M, Fisher I (2008) Phys Rev B 77(3):035114
13. Hamlin J, Zocco D, Sayles T, Maple M, Chu JH, Fisher I (2009) Phys Rev Lett 102(17):177002
14. Kusmartseva AF, Sipos B, Berger H, Forro L, Tutiš E (2009) Phys Rev Lett 103(23):236401
15. Kim H, Kang CJ, Kim K, Shim JH, Min BI (2015) Phys Rev B 91:165130
16. Magdysyuk OV, Nuss J, Jansen M (2013) Acta Crystallogr Sect B Struct Sci Cryst Eng Mater 69(6):547
17. Nuss J, Jansen M (2009) J Alloys Compd 480(1):57
18. Wang PL, Kolodiazhnyi T, Yao J, Mozharivskyj Y (2012) J Am Chem Soc 134(3):1426
19. Kim H, Kang CJ, Kim K, Shim J, Min B (2016) Phys Rev B 93(12):125116
20. Abrikosov AA (1994) Fundamentals of the theory of metals. Yoshiokashoten
21. Jung MH, Lacerda AH, Takabatake T (2002) Phys Rev B 65:132405
22. Han F, Wan X, Phelan D, Stoumpos CC, Sturza M, Malliakas CD, Li Q, Han TH, Zhao Q, Chung DY, Kanatzidis MG (2015) Phys Rev B 92:045112
23. Kamihara Y, Watanabe T, Hirano M, Hosono H (2008) J Am Chem Soc 130(11):3296
24. Mizuguchi Y, Demura S, Deguchi K, Takano Y, Fujihisa H, Gotoh Y, Izawa H, Miura O (2012) J Phys Soc Jpn 81(11):114725
25. Matthias BT, Corenzwit E (1957) Phys Rev 107:1558
26. Winiarski MJ, Wiendlocha B, Gołąb S, Kushwaha SK, Wiśniewski P, Kaczorowski D, Thompson JD, Cava RJ, Klimczuk T (2016) Phys Chem Chem Phys 18(31):21737
27. Sturza M, Han F, Malliakas CD, Chung DY, Claus H, Kanatzidis MG (2014) Phys Rev B 89:054512
28. Landau LD, Ginzburg V (1950) Zh Eksp Teor Fiz 20:1064
29. Ghiringhelli G, Le Tacon M, Minola M, Blanco-Canosa S, Mazzoli C, Brookes N, De Luca G, Frano A, Hawthorn D, He F et al (2012) Science:1223532
30. Chang J, Blackburn E, Holmes A, Christensen N, Larsen J, Mesot J, Liang R, Bonn D, Hardy W, Watenphul A et al (2012) Nat Phys 8(12):871
31. Johnston DC (2010) Adv Phys 59(6):803
32. Hosono H, Tanabe K, Takayama-Muromachi E, Kageyama H, Yamanaka S, Kumakura H, Nohara M, Hiramatsu H, Fujitsu S (2016) Sci Technol Adv Mater
33. Lee G, Farhan MA, Kim JS, Shim JH (2013) Phys Rev B 87(24):245104
34. Feng Y, Wang Z, Chen C, Shi Y, Xie Z, Yi H, Liang A, He S, He J, Peng Y et al (2014) Sci Rep 4:5385
35. Kodama K, Wakimoto S, Igawa N, Shamoto S, Mizoguchi H, Hosono H (2011) Phys Rev B 83(21):214512

Chapter 6
Universal Superconductivity in R_2O_2Bi

6.1 Introduction

Two-dimensional superconductivity of Bi^{2-} square net in Y_2O_2Bi and Tb_2O_2Bi was realized for the first time by enhancing two-dimensionality via O incorporation. In addition, Tb_2O_2Bi showed novel electronic properties as were not seen in other R_2O_2Bi compounds. These results indicate that R_2O_2Bi is the new platform for two-dimensional superconductivity with the emergence of novel electronic phenomena by changing R^{3+}.

In this study, I systematically synthesized R_2O_2Bi (R: Y, Tb, Dy, and Lu) samples. All compounds incorporated O into their frameworks, resulting in the emergence of superconductivity under a universal rule as described below.

6.2 Experimental

Polycrystalline Dy_2O_2Bi and Lu_2O_2Bi samples were additionally synthesized by the same procedure to that of Tb_2O_2Bi. Detailed conditions were summarized in Appendices B and C. Crystal structures were evaluated by powder XRD using Cu Kα radiation. Rietveld refinements were performed by RIETAN-FP [1]. Electrical transport properties were evaluated by the standard four-probe method.

6.3 Results and Discussion

Mixing CaO during synthesis as a O supplier was used to expand c-axis length via O incorporation for Dy_2O_2Bi and Lu_2O_2Bi. The reason why only Y_2O_2Bi achieved O incorporation without CaO was probably attributed to the easier formation of

R. Sei, *Two-Dimensional Superconductivity in Rare Earth Oxybismuthides with Unusual Valent Bismuth Square Net*, Springer Theses,
https://doi.org/10.1007/978-981-99-7313-2_6

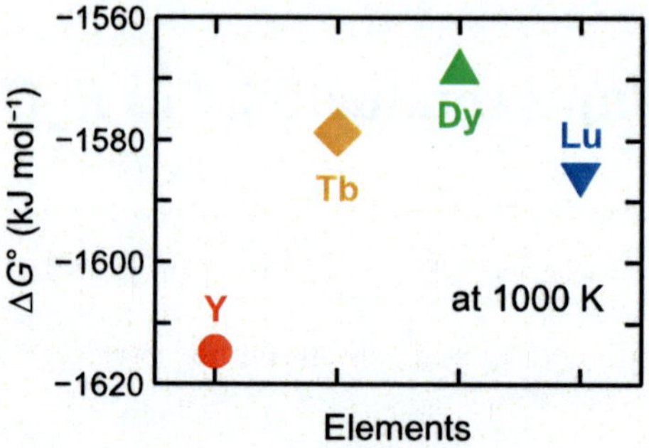

Fig. 6.1 Gibbs free energy of formation for various R_2O_3 at 1000 K

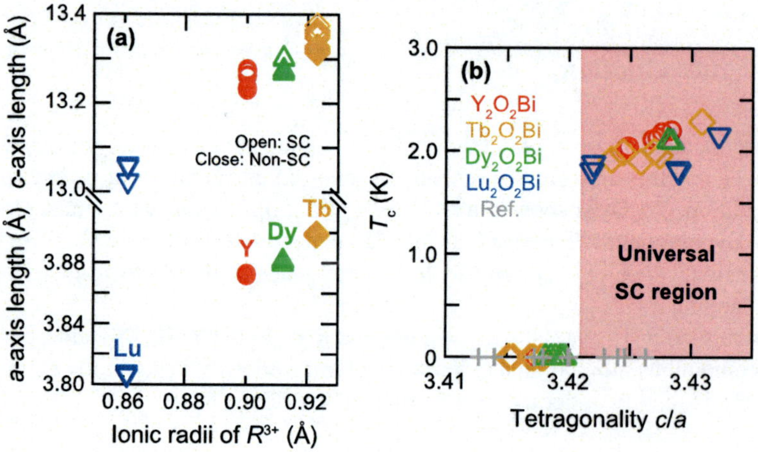

Fig. 6.2 **a** Lattice constant variation and **b** superconducting phase diagram of R_2O_2Bi compounds. SC: superconducting

Y–O bonding than those of other R, supported by the Gibbs free energy of formation for R_2O_3 at 1000 K (Fig. 6.1) [2]. Figure 6.2a shows the lattice constant variation of R_2O_2Bi compounds plotted as a function of ionic radii of R^{3+} [3]. Both a and c-axes varied according to the lanthanoid contraction rule of R^{3+} consistent with the previous study [4], supporting their trivalent state. Each R_2O_2Bi showed c-axis expansion via O incorporation with almost constant a-axis length connecting to the emergence of the superconductivity, except for Lu_2O_2Bi: all Lu_2O_2Bi samples showed the onset of the superconductivity (Appendix C). Generally, T_c in each R_2O_2Bi was proportional to the c-axis length. However, there were no universal relation between T_c and c-axis length among all R_2O_2Bi compounds.

Such an obscure situation could be resolved by introducing a new parameter called tetragonality, i.e., c/a, which was a structural aspect ratio connected with the two-dimensionality of Bi^{2-} square net in each R_2O_2Bi. Figure 6.2b shows the superconducting phase diagram as a function of the c/a in R_2O_2Bi compounds.

Notably, the superconductivity of all R_2O_2Bi was universally observed for $c/a > 3.42$ with few exceptions in previous studies [4, 5]. This result strongly suggests that the superconductivity was preferentially emerged in isolated monatomic Bi^{2-} square net as was seen in Y_2O_2Bi and Tb_2O_2Bi.

Figure 6.3 shows a schematic image of the rich electronic phases of Bi^{2-} square net observed in this study. The pristine R_2O_2Bi showed normal metallic behavior without

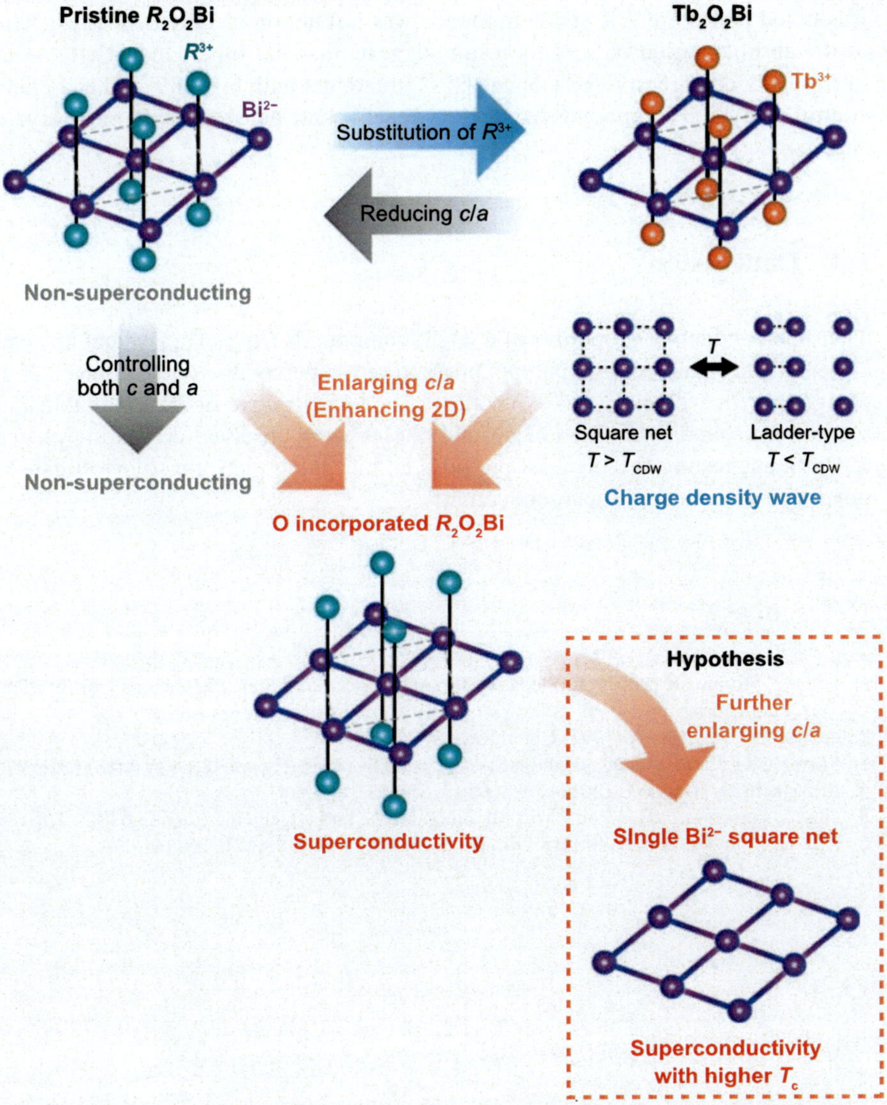

Fig. 6.3 A schematic image of rich physical properties of Bi^{2-} square net in R_2O_2Bi

the superconductivity down to 1.8 K, corresponding to the previous report [4]. This property was maintained for various R^{3+} with constant c/a except for Tb_2O_2Bi, which only showed the CDW transition probably due to the weakened spin–orbit coupling [6]. The CDW transition was suppressed by applying uniaxial chemical pressure along c-axis in which reducing c/a resulted in the non-appearance of the superconductivity and increasing c/a resulted in the emergence of the superconductivity. The superconductivity universally emerged in all R_2O_2Bi compounds by enhancing the two-dimensionality of Bi^{2-} square net via the chemical control of c/a. It is noted that T_c of R_2O_2Bi compounds was a function of only c/a, irrespective of the absolute value of a. This result suggests that the superconductivity would intrinsically emerge in more isolated Bi^{2-} square net with higher T_c. The chemical control of layered compounds like this study will contribute to finding novel layered superconductors.

6.4 Conclusion

The superconducting properties of R_2O_2Bi compounds (R: Y, Tb, Dy, and Lu) were systematically studied. All compounds showed superconductivity around 2 K by enlarging c/a, where T_c was a function of c/a irrespective of R^{3+}. The threshold value was around 3.42. Rich electronic phases were observed depending on R in R_2O_2Bi, suggesting that R_2O_2Bi provides an intriguing platform to investigate the novel two-dimensional superconductivity.

References

1. Izumi F, Momma K (2007) Applied crystallography XX, solid State Phenomena, vol 130. Trans Tech Publications, pp 15–20. https://doi.org/10.4028/www.scientific.net/SSP.130.15
2. Adachi GY, Imanaka N (1998) Chem Rev 98(4):1479
3. Shannon RT (1976) Acta Crystallogr Sect A Cryst Phys Diffr Theoret Gen Crystallogr 32(5):751
4. Mizoguchi H, Hosono H (2011) J Am Chem Soc 133(8):2394
5. Nuss J, Jansen M (2012) Zeitschrift für anorganische und allgemeine Chemie 638(3–4):611
6. Kim H, Kang CJ, Kim K, Shim J, Min B (2016) Phys Rev B 93(12):125116

Chapter 7
General Conclusion

In this thesis, I developed novel synthetic routes for both epitaxial thin films and polycrystalline powders of R_2O_2Bi compounds in order to explore the detailed physical properties of unusual Bi^{2-} square net. These approaches enabled to discover and characterize the different fascinating phenomena from those of well-studied Bi^{-} square net in $AMBi_2$ compounds.

In the thin film study, I succeeded in obtaining Y_2O_2Bi epitaxial thin film for the first time by developing two SPE methods. Reductive SPE utilized the direct reaction between (Y, Bi) powders and Y_2O_3 amorphous thin film. This method indicates that the preferential reduction of Bi by strong reductant Y was essential to achieve Y_2O_2Bi phase, although surface residues hampered the fundamental evaluation of physical properties. Therefore, this concept was applied to the improved multilayer SPE method where all reaction precursors changed to thin film form, enabling the fabrication of high-quality epitaxial thin films with a flat surface and without any residues. The multilayer SPE method could be universally applied to obtain other R_2O_2Bi epitaxial thin films such as Ce_2O_2Bi and Dy_2O_2Bi, which will accelerate the study about Bi^{2-} square net in future. The unusual reductive state of Bi^{2-} square net was directly confirmed in Y_2O_2Bi epitaxial thin film by XPS measurement. Furthermore, a strong interaction between two-dimensional electronic nature and strong spin–orbit coupling in Bi^{2-} square net was observed in electrical transport properties, demonstrating the possible emergence of topological insulating behavior in future study.

In the study about polycrystalline powders, I discovered two-dimensional superconductivity of Bi^{2-} square net in R_2O_2Bi compounds which was concluded as non-superconducting in previous studies. This superconductivity was universally induced in R_2O_2Bi compounds by enlarging c/a via O incorporation, which corresponded to the enhancement of two-dimensionality of Bi^{2-} square net. Furthermore, rich electronic phases appeared depending on R^{3+} in R_2O_2Bi, including the CDW instability whose presence contradicted the previous theoretical report. The competition among

R. Sei, *Two-Dimensional Superconductivity in Rare Earth Oxybismuthides with Unusual Valent Bismuth Square Net*, Springer Theses,
https://doi.org/10.1007/978-981-99-7313-2_7

superconductivity, AF ordering, and CDW in Tb_2O_2Bi suggested the unconventional superconductivity of Bi^{2-} square net. These phenomena were possibly originating from the different influence of each R^{3+} on Bi^{2-} square net. Therefore, R^{3+} substitution and c/a control were crucial to engineer the properties of Bi^{2-} square net, and R_2O_2Bi would provide an intriguing platform to investigate the interplay between two-dimensional superconductivity and strong spin–orbit coupling.

The main findings in this study were the discovery of novel properties (e.g., superconductivity and two-dimensional electronic nature with strong spin–orbit coupling) and the establishment of exploring techniques (e.g., novel SPE methods and chemical control of c/a via O incorporation by CaO mixing and F substitution). The latter would be applicable to not only R_2O_2Bi compounds but also other layered compounds, probably shedding the light on hidden layered compounds. Establishment of these concepts would be the most valuable result in this study because they would contribute to the progress of rich physics in layered compounds.

Appendix A
Post-Growth of Superconducting Y_2O_2Bi

A.1 Introduction

The superconductivity in Y_2O_2Bi was induced by enlarging c/a via the increase of nominal amount of O in starting materials. However, this simple synthesis procedure resulted in the increase of Y_2O_3 amount as an impurity phase. Furthermore, the mechanism of enlarged c/a was not confirmed yet; the O incorporation was one reasonable assumption. Therefore, the improvement of phase purity and the reinforcement of O incorporation were very meaningful for application to other systems and future development.

When only Y_2O_3 was heated in an evacuated quartz tube, the color changed from white to gray. This implied the generation of O vacancy from Y_2O_3 impurity during the synthesis of Y_2O_2Bi and the released O would be incorporated into Y_2O_2Bi framework. According to this situation, the O incorporation would be also achieved by post reaction between pristine Y_2O_2Bi and generated O from Y_2O_3. This appendix verified the post growth and reversibility of superconducting Y_2O_2Bi in order to improve the phase purity and to reinforce the O incorporation scenario.

A.2 Experimental

Four Y_2O_2Bi samples A–D were newly synthesized by the following procedures:

A. Trial to synthesize pure Y_2O_2Bi sample by a similar procedure to that of $x = 1.3$ in Sect. 4.2.
B. Heating sample A with Y_2O_3 pellet (same mole) in evacuated quartz tube at 1000°C for 10 h. Only Y_2O_2Bi pellet was covered with Ta foil.

R. Sei, *Two-Dimensional Superconductivity in Rare Earth Oxybismuthides with Unusual Valent Bismuth Square Net*, Springer Theses,
https://doi.org/10.1007/978-981-99-7313-2

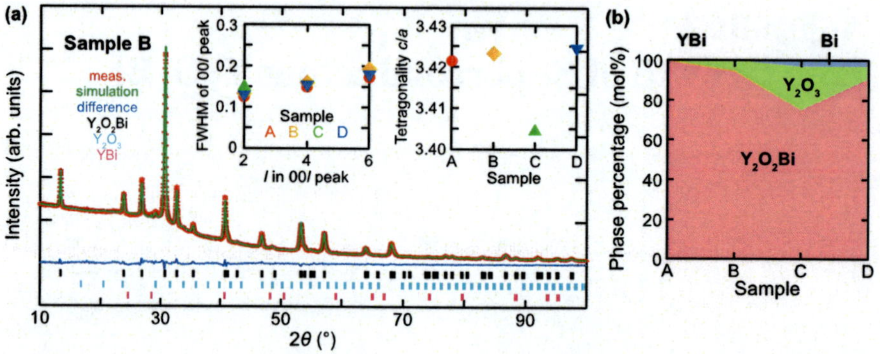

Fig. A.1 **a** Powder XRD pattern of sample B with Rietveld refinement. Insets show the FWHM of $00l$ peaks in 2θ-θ pattern (left) and variation of c/a (right). **b** Phase diagram for Y_2O_2Bi samples

C. Heating sample B with Y pellet (same mole) in evacuated quartz tube at 1000°C for 10 h. Y_2O_2Bi pellet and Y pellet were covered with Ta foil together in order to avoid the reaction between Y and quartz.
D. Heating only sample B in evacuated quartz tube at 1000°C for 10 h.

A.3 Results and Discussion

Figure A.1a shows a powder XRD pattern of sample B. Most of the diffraction peaks were fitted by the simulation of $ThCr_2Si_2$ -type Y_2O_2Bi. This situation was the same for all samples A–D, indicating that the Y_2O_2Bi framework was retained in any experimental operations. The samples A, B, and D showed high purity of Y_2O_2Bi phase. On the other hand, sample C showed the increase of impurity phases probably due to the drastic reactivity of Y suggesting that the milder reductant was better.[1] Notably, the color of Y_2O_3 pellet changed from white to gray after the synthesis of sample B followed by the return to white by exposing to air for several days, indicating the generation of O vacancy in Y_2O_3 during the experiment as expected.

The structural parameters for samples A–D are summarized in Table A.1. The c/a values changed reversibly by performing post oxidation (sample B) and reduction (sample C). This strongly supported the O incorporation scenario because only O amount was controllable through these experimental operations. The FWHM of $00l$ diffraction peaks in 2θ-θ pattern showed similar values among all samples, indicating that the post O incorporation and removal proceeded on not only the pellet surface but also its inside. The sample D still showed large c/a, suggesting that incorporated O formed a chemical bonding to $[Y_2O_2]^{2+}$ slab. The reason why sample A showed larger c/a than the sample for $x = 1.3$ in Chap. 4 was attributed to the oxidation of

[1] The large S in Rietveld refinement for sample C (Table A.1) was attributed to the broad peak around $2\theta = 23°$ originating from the double-sided tape (Appendix C.3).

Table A.1 Summary of structural data of Y_2O_2Bi in this appendix

Sample	A	B	C	D
a (Å)	3.87281(14)	3.87397(17)	3.8733(3)	3.87370(16)
c (Å)	13.2503(5)	13.2617(5)	13.1864(13)	13.2654(5)
c/a	3.42138(17)	3.4233(2)	3.4044(4)	3.4245(2)
V (Å^3)	198.738(13)	199.026(15)	198.34(3)	199.055(14)
Y z	0.33097(10)	0.32965(10)	0.3235(3)	0.33028(10)
Purity (mol%)	99.3	94.1	75.0	91.1
R_{wp}	2.260	2.269	2.725	2.354
R_e	1.342	1.337	1.315	1.320
S	1.6849	1.6970	2.0715	1.7829

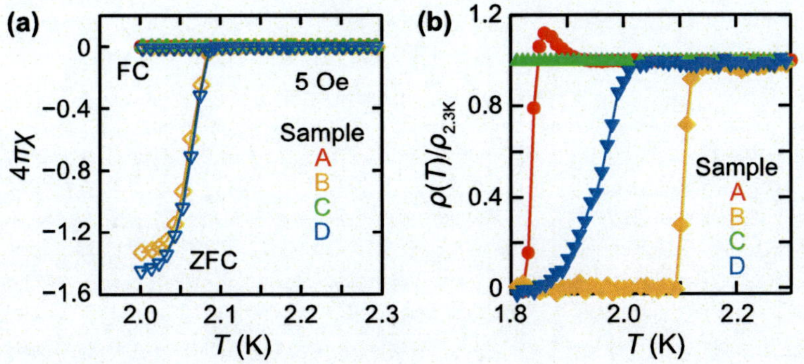

Fig. A.2 **a** The χ-T curves for samples A–D with ZFC and FC process at 5 Oe. **b** The normalized ρ-T curves near T_c for samples A–D

starting material Y due to the movement in the air for the relocation of experimental laboratory.

According to the c/a values, samples A, B, and D were expected to exhibit the superconductivity ($c/a > 3.42$). Figure A.2 shows χ-T and ρ-T curves for all samples. The superconductivity was observed in samples A, B, and D with T_c derived from ρ-T curves of 1.85, 2.12, and 2.00 K, respectively, which was consistent with the result in Chap. 6. Sample D showed a broad transition in ρ-T curves suggesting the partial volume fraction of the superconductivity due to the reduction by vacuum heating, although sample D showed a large diamagnetic signal and similar FWHM to other samples.

Here, I focus on the superconductivity of sample B whose properties were enhanced by post O incorporation from sample A. This sample showed bulk superconductivity with same T_c to the sample for $x = 1.7$ in Chap. 4. On the other hand, the purity of Y_2O_2Bi phase was greatly improved from 70.9 to 94.1mol%. This indicates that the post O incorporation method in this appendix was far superior to the previous method which just increased the nominal amount of O in starting materials.

A.4 Conclusion

The post control method of the c/a in Y_2O_2Bi was newly established. This enabled to synthesize the high-purity superconducting samples, in addition to the reversible engineering between superconducting and non-superconducting samples by performing post oxidation and reduction. These results reinforced the O incorporation scenario in Y_2O_2Bi. By optimizing Y_2O_3 amount or O source, high-quality superconducting Y_2O_2Bi with higher T_c and purity would become available.

Appendix B
Synthesis of Polycrystalline Dy_2O_2Bi

B.1 Introduction

The $ThCr_2Si_2$ -type Dy_2O_2Bi is already synthesized, but the physical properties are not investigated as that as the case of Tb_2O_2Bi [1]. The Dy^{3+} has similar ionic radii to Y^{3+}, so it is expected that the difference in physical properties from Y_2O_2Bi is influenced mainly by magnetism (Dy^{3+}: $4f^9$) with a small contribution of chemical pressure. In this appendix, polycrystalline Dy_2O_2Bi samples were successfully synthesized and their physical properties were investigated for the first time.

B.2 Experimental

Pristine, F doped, and O incorporated polycrystalline $ThCr_2Si_2$-type Dy_2O_2Bi samples were synthesized by the same procedure to that of Tb_2O_2Bi, although the second pelletizing process was conducted in air. The starting materials were Bi (3N), Dy (3N), Dy_2O_3 (3N), BiF_3 (3N), and CaO prepared by heating $CaCO_3$ (3Nup). The nominal composition of starting materials is summarized in Table B.1. Crystal structures were evaluated by powder XRD using Cu Kα radiation. Rietveld refinements were performed by RIETAN-FP [2]. Magnetic properties were measured by SQUID magnetometer. Electrical transport properties were evaluated by the standard four-probe method. Specific heat measurements were conducted by the thermal relaxation method.

R. Sei, *Two-Dimensional Superconductivity in Rare Earth Oxybismuthides with Unusual Valent Bismuth Square Net*, Springer Theses,
https://doi.org/10.1007/978-981-99-7313-2

Table B.1 Summary of structural data in pristine Dy_2O_2Bi samples

Nominal O amount x	1.3	1.4	1.5	1.6	1.7	1.8	1.9
a (Å)	3.8824(2)	3.8820(3)	3.8809(7)	3.8819(8)	3.8822(3)	3.8821(2)	3.8819(2)
c (Å)	13.2727(6)	13.2695(9)	13.269(2)	13.274(3)	13.2758(11)	13.2753(8)	13.2742(7)
c/a	3.4186(3)	3.4182(3)	3.4191(9)	3.4194(10)	3.4197(3)	3.4197(3)	3.4195(3)
V (Å^3)	200.64(19)	199.97(3)	199.85(2)	200.03(7)	200.08(3)	200.07(2)	200.032(19)
Dy z	0.3287(2)	0.3284(3)	0.3276(2)	0.3290(6)	0.3326(5)	0.3319(5)	0.3322(6)
Purity (mol%)	99.1	>99	82.3	74.0	63.6	62.8	50.5
R_{wp}	1.589	1.406	1.247	1.249	1.181	1.101	1.069
R_e	0.945	0.943	0.911	0.884	0.883	0.884	0.863
S	1.6828	1.4917	1.3679	1.4129	1.3368	1.2453	1.2390

B.3 Results and Discussion for Pristine Dy_2O_2Bi

Figure B.1a shows a typical powder XRD pattern of Dy_2O_2Bi sample for $x = 1.4$. Most of the diffraction peaks were completely fitted by simulation as $ThCr_2Si_2$-type Dy_2O_2Bi phase, indicating successful synthesis. Refined structural parameters and phase purity are summarized in Table B.1 and Fig. B.1b. The lattice constants were similar to those in the previous study ($a = 3.8761$, $c = 13.233$ Å) [1]. AF insulator Dy_2O_3 ($T_N = 1.2$ K) [3] appeared as an impurity in most of samples, although there was no influence on the physical properties of Dy_2O_2Bi samples in this study. The c/a was almost constant regardless of x (inset of Fig. B.1 and Table B.1) with a smaller value than the superconducting threshold value 3.42, similar to the result of Tb_2O_2Bi. This result indicated the lack of O incorporation in Dy_2O_2Bi and absence of superconductivity.

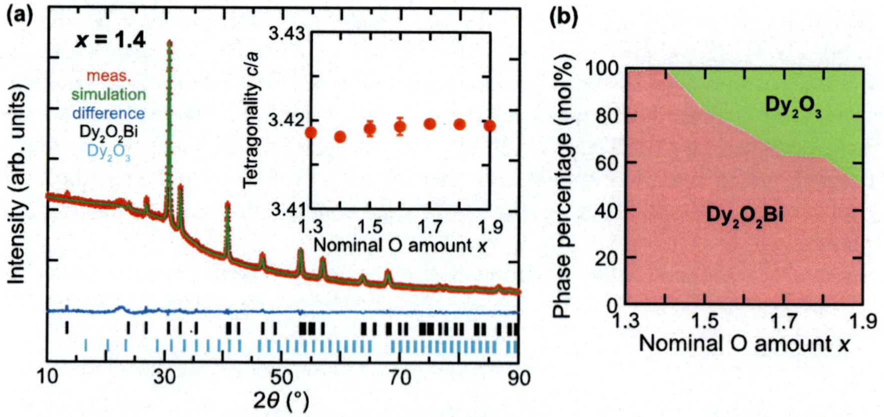

Fig. B.1 **a** Typical powder XRD pattern of Dy_2O_2Bi for $x = 1.4$ with Rietveld refinement. Inset shows the variation of c/a. **b** Phase diagram for Dy_2O_2Bi samples

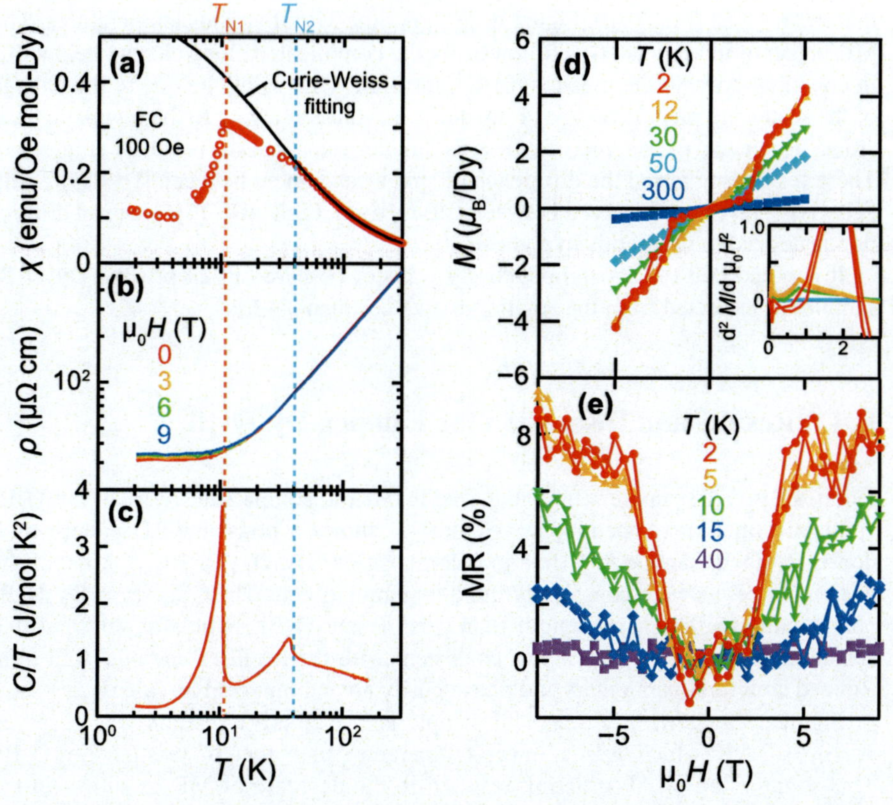

Fig. B.2 Physical properties of Dy_2O_2Bi for $x = 1.4$. Temperature dependence of **a** χ, **b** ρ, and **c** C/T. Broken lines denote first and second Néel temperatures, T_{N1} and T_{N2}. (d) Magnetization curve with second derivative (inset) and (e) MR at various temperatures

Figure B.2a shows χ-T curve of Dy_2O_2Bi sample for $x = 1.4$. A sudden χ drop at $T_{N1} = 10.8$ K indicated the AF transition of Dy_2O_2 layer, as observed in other magnetic R_2O_2Bi (Table 5.2). This transition was also observed in ρ-T curve and T dependence of C/T: small ρ drop (Fig. B.2b) and λ-shape peak of C/T (Fig. B.2c). Notably, χ-T curve showed a small deviation from the Curie–Weiss fitting around 40 K as was never seen in R_2O_2Bi compounds. The small λ-shape peak in C/T versus T plot at the same temperature strongly indicated the existence of AF ordering even at $T_{N2} = 40$ K. Such multistep magnetic transitions were sometimes observed in rare earth compounds [4]. The effective Bohr magneton derived from Curie–Weiss fitting ($40 < T < 300$ K) was $10.86\mu_B$ which was similar to theoretical value $10.64\mu_B$ for Dy^{3+}, confirming the trivalent state of Dy.

Figure B.2d shows the magnetization curve at various temperatures. Metamagnetic transition was observed below T_{N2} at 1–2 T as shown in the inset of Fig. B.2. Furthermore, the increase of M became sharper at 2 K (below T_{N1}) compared with that in Tb_2O_2Bi (Fig. 5.3d). Figure B.2e shows the MR at various temperatures. The MR began to increase at T_{N2}, followed by the complicated slope change below T_{N1} in combination with the metamagnetic transition. This would be due to the change of the scattering process and/or Fermi surface evolution owing to the magnetic transition. However, the degree of MR was smaller than that of Tb_2O_2Bi (Fig. 5.3e). These results suggested the difference of spin arrangement between $[Dy_2O_2]^{2+}$ and $[Tb_2O_2]^{2+}$ layers as observed between $CaMnBi_2$ and $SrMnBi_2$ [5]. Neutron diffraction is desired to verify this hypothesis.

It is noted that the superconductivity was not observed in all pristine Dy_2O_2Bi samples as expected from the smaller c/a values than 3.42.

B.4 Results and Discussion for F doped Dy_2O_2Bi

F doped Dy_2O_2Bi sample was synthesized in order to compare the result of Dy_2O_2Bi epitaxial thin film (Appendix D). Figure B.3 shows a powder XRD pattern of F doped Dy_2O_2Bi sample #F1 (nominal composition: $Dy_2O_{1.2}F_{0.2}Bi_{1.6}$). Most of the diffraction peaks were completely fitted by simulation as $ThCr_2Si_2$ -type Dy_2O_2Bi phase with smaller c-axis length than pristine Dy_2O_2Bi, indicating successful F substitution for O site, which was consistent with the result of epitaxial thin film. Refined structural parameters and phase purity are summarized in Table B.2.

Figure B.4a shows χ-T curve of sample #F1. The AF transition was observed at $T_{N1} = 7.5$ K which was a lower temperature than that of pristine Dy_2O_2Bi. This suppression of AF ordering resulted in the decrease of MR as shown in the inset of Fig. B.4. These results were consistent with those of the epitaxial thin film (Appendix D). The derived effective Bohr magneton value was $10.94\mu_B$, confirming the Dy^{3+} state.

In contrast to the results of pristine Dy_2O_2Bi, there were no clear deviations of χ-T curve from Curie–Weiss fitting. However, by deriving the second derivative of χ-T curve, the possible existence of magnetic ordering around 30 K was suggested similar to the result of pristine Dy_2O_2Bi (inset of Fig. B.4a). In order to verify this anomaly, specific heat measurement was conducted. Figure B.4c shows the T dependence of C/T for both F doped and pristine Dy_2O_2Bi samples. The decrease of T_{N1} was

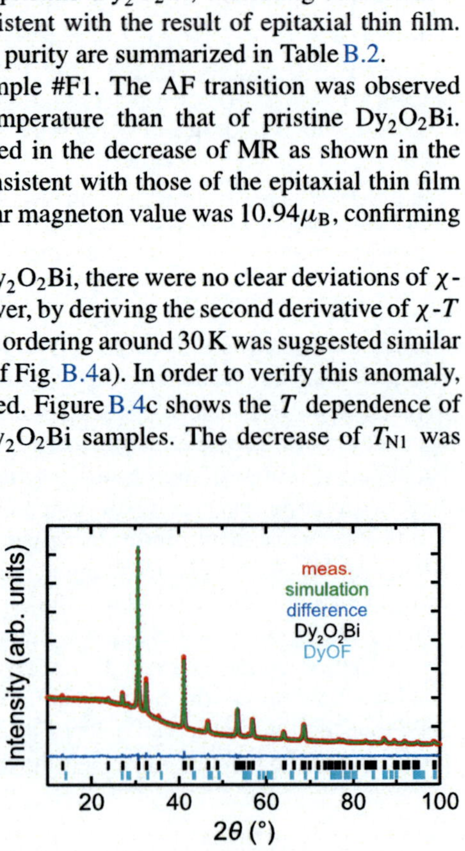

Fig. B.3 Powder XRD pattern of F doped Dy_2O_2Bi with Rietveld refinement

Table B.2 Summary of structural data in F doped and O incorporated Dy_2O_2Bi samples

Sample #	F1	SC1	SC2
Nominal composition	$Dy_2O_{1.2}F_{0.2}Bi_{1.6}$	$Dy_2O_{2.0}Bi_{1.2}$ + $(CaO)_{0.5}$	$Dy_2O_{2.0}Bi_{1.0}$ + $(CaO)_{0.5}$
a (Å)	3.88785(13)	3.8807(3)	3.8791(3)
c (Å)	13.1324(4)	13.3035(15)	13.2990(11)
c/a	3.37781(15)	3.4281(5)	3.4284(4)
V (Å^3)	198.501(11)	200.35(3)	200.11(3)
Dy z	0.32985(18)	0.3222(4)	0.3273(7)
Purity (mol%)	90.3	26.3	43.7
R_{wp}	1.477	1.307	1.193
R_e	0.990	0.826	0.790
S	1.4928	1.5818	1.5092

clearly observed. Furthermore, a broad slope change was observed around 30 K for F doped sample. This suggested the existence of the weak magnetic ordering around this temperature.

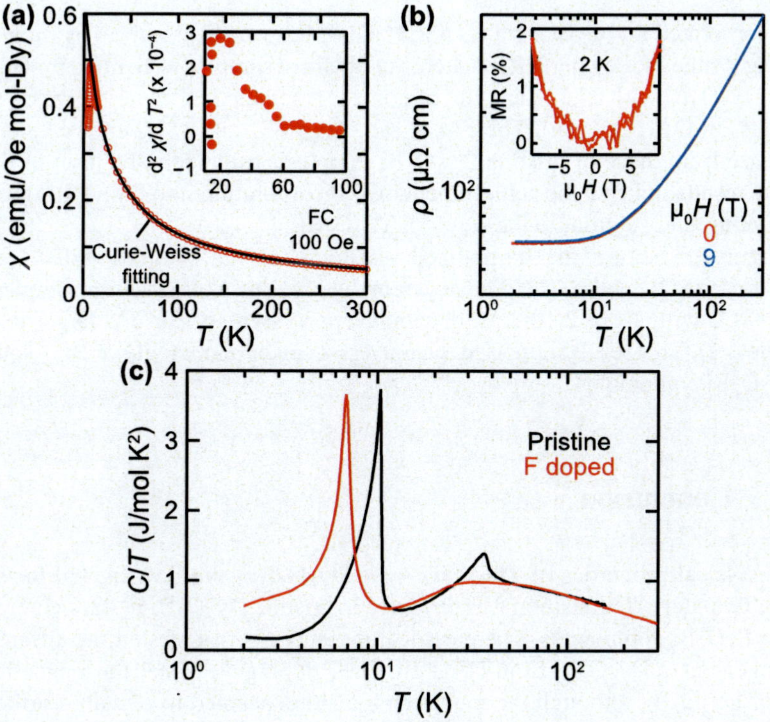

Fig. B.4 Physical properties of F doped Dy_2O_2Bi sample #F1. **a** χ-T curve and its second derivative (inset). **b** ρ-T curve and MR data at 2 K. (c) Temperature dependence of C/T for both F doped and pristine Dy_2O_2Bi samples

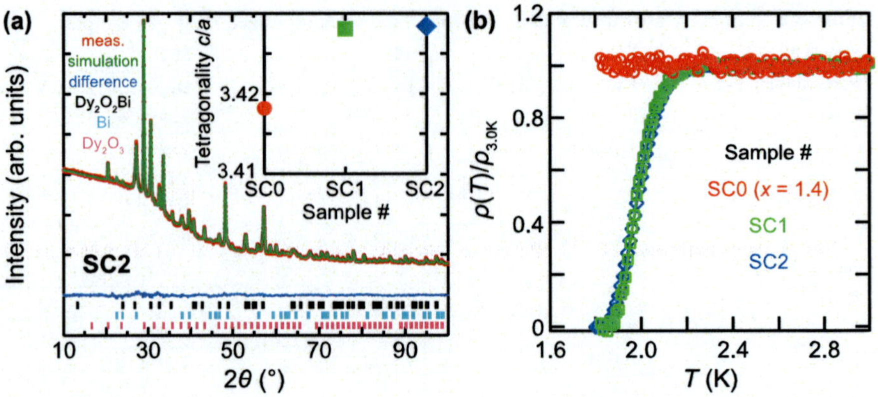

Fig. B.5 **a** Powder XRD patterns of O incorporated Dy_2O_2Bi for sample #SC2 with Rietveld refinement. Inset shows the variation of c/a. SC0 is the result for $x = 1.4$. **b** Normalized ρ-T curve for O incorporated Dy_2O_2Bi samples

B.5 Results and Discussion for O Incorporated Dy_2O_2Bi

O incorporated Dy_2O_2Bi samples #SC1 and #SC2 were synthesized by mixing CaO during synthesis. Powder XRD patterns of obtained samples were fitted by $ThCr_2Si_2$-type Dy_2O_2Bi with a large amount of non-superconducting impurity phases such as Bi and Dy_2O_3 (Fig. B.5a). The c/a of samples #SC1 and #SC2 showed a dramatic increase from those of pristine Dy_2O_2Bi samples as shown in the inset of Fig. B.5. These results indicate the achievement of O incorporation into Dy_2O_2Bi framework by mixing CaO, similar to the result of Tb_2O_2Bi.

Figure B.5 shows the normalized resistivity of the samples #SC0 (equal to $x = 1.4$), #SC1, and #SC2. The superconductivity was observed in samples #SC1 and #SC2 with T_c of 2.10K. Both samples possessed similar c/a larger than 3.42 (Table B.2). This result supported the universal superconductivity of Bi^{2-} square net in R_2O_2Bi compounds.

B.6 Conclusion

The physical properties of $ThCr_2Si_2$ -type Dy_2O_2Bi were investigated for the first time. Two-step AF transition was observed in pristine Dy_2O_2Bi in stark contrast with other R_2O_2Bi compounds. The electrical measurement suggested the different spin arrangement of Dy_2O_2Bi from Tb_2O_2Bi. The F doping weakened the magnetic ordering of Dy_2O_2Bi, although the multistep transition seemed to remain. Furthermore, the superconductivity was also observed by enlarging c/a, supporting the universal superconductivity of Bi^{2-} square net in R_2O_2Bi.

References

1. Nuss J, Jansen M (2012) Zeitschrift für anorganische und allgemeine Chemie 638(3–4):611
2. Izumi F, Momma K (2007) Applied crystallography XX, solid state phenomena, vol 130, pp 15–20. Trans Tech Publications. https://doi.org/10.4028/www.scientific.net/SSP.130.15
3. Adachi GY, Imanaka N (1998) Chem Rev 98(4):1479
4. Gignoux D, Schmitt D (1991) J Magn Magn Mater 100(1–3):99
5. Guo Y, Princep A, Zhang X, Manuel P, Khalyavin D, Mazin I, Shi Y, Boothroyd A (2014) Phys Rev B 90(7):075120

Appendix C
Synthesis of Polycrystalline Lu_2O_2Bi

C.1 Introduction

There are no reports about the synthesis of $ThCr_2Si_2$ -type Lu_2O_2Bi phase. The ionic radius of Lu^{3+} is quite smaller than other R^{3+} cations in this thesis, possibly resulting in the deformation of the crystal structure although CDW instability should be suppressed by the strong chemical pressure effect [1]. This appendix shows the first report of synthesis and electrical transport properties of $ThCr_2Si_2$ -type Lu_2O_2Bi phase.

C.2 Experimental

Pristine and O incorporated polycrystalline Lu_2O_2Bi samples were synthesized by the same procedure to that of Tb_2O_2Bi. The starting materials were Bi (3N), Lu (3N), Lu_2O_3 (3N), and CaO prepared by heating $CaCO_3$ (3Nup). The nominal composition of starting materials is summarized in Table C.1. Crystal structures were evaluated by powder XRD using Cu Kα radiation. Rietveld refinements were performed by RIETAN-FP [2]. Electrical transport properties were evaluated by the standard four-probe method.

C.3 Results and Discussion

Figure C.1a shows typical powder XRD patterns of Lu_2O_2Bi sample #3, which possessed highest purity (99.3mol%). Most of the diffraction peaks were fitted by the simulation of $ThCr_2Si_2$ -type structure, indicating that Lu_2O_2Bi also possessed the

R. Sei, *Two-Dimensional Superconductivity in Rare Earth Oxybismuthides with Unusual Valent Bismuth Square Net*, Springer Theses,
https://doi.org/10.1007/978-981-99-7313-2

Table C.1 Summary of structural data in Lu_2O_2Bi samples

Sample #	1	2	3	4	5
Nominal composition	$Lu_2O_{0.7}Bi_{2.0}$	$Lu_2O_{1.0}Bi_{2.0}$	$Lu_2O_{1.3}Bi_{1.7}$	$Lu_2O_{1.6}Bi_{1.4}$	$Lu_2O_{2.0}Bi_{1.0}$ + $(CaO)_{0.5}$
Space group	*I4/mmm*	*I4/mmm*	*I4/mmm*	*I4/mmm*	*I4/mmm*
a (Å)	3.8057(4)	3.8065(5)	3.8077(3)	3.8084(4)	3.8061(3)
c (Å)	13.0226(12)	13.0259(16)	13.0560(9)	13.0590(12)	13.0633(7)
c/a	3.4218(5)	3.4220(6)	3.4289(3)	3.4290(4)	3.4322(3)
V (Å^3)	188.61(3)	188.74(4)	189.29(2)	189.41(3)	189.24(2)
Lu *z*	0.3338(4)	0.3338(4)	0.3294(3)	0.3271(3)	0.3251(4)
Purity (mol%)	70.6	81.0	99.3	83.7	19.3
R_{wp}	2.129	2.195	2.645	2.134	1.947
R_e	1.146	1.310	1.097	0.941	0.941
S	1.8575	1.6754	2.4108	2.2667	2.0677

Wyckoff positions: Bi on the 2*a* site (0, 0, 0), Lu on the 4*e* site (0, 0, *z*), and O on the 4*d* site (0, 0.5, 0.25)

same crystal structure as other R_2O_2Bi. This was the first synthesis of $ThCr_2Si_2$-type Lu_2O_2Bi. The broad peak observed around $2\theta = 23°$ was originating from the double-sided tape which was used to fix the sample, and it was the main cause for a large S value in the Rietveld refinement. The lattice constants are summarized in Table C.1 and the trend of impurity phases was similar to that of Tb_2O_2Bi (Fig. C.1b).

Inset of Fig. C.1a shows the variation of c/a in Lu_2O_2Bi samples. There are three groups of c/a values: samples #1 and #2 were small, #3 and #4 were middle, and #5 was a large value. The samples #1 and #2 were synthesized in a reductive condition, indicating that the decrease of c-axis length was originating from O vacancy. The sample #3 was almost pure while further oxidative synthesis (#4) resulted in no change in the lattice constants, indicating their ideal composition. Therefore, only sample #5 was the O incorporated Lu_2O_2Bi. These results indicate that mixing CaO was also effective for Lu_2O_2Bi to achieve O incorporation. Notably, the c/a of all

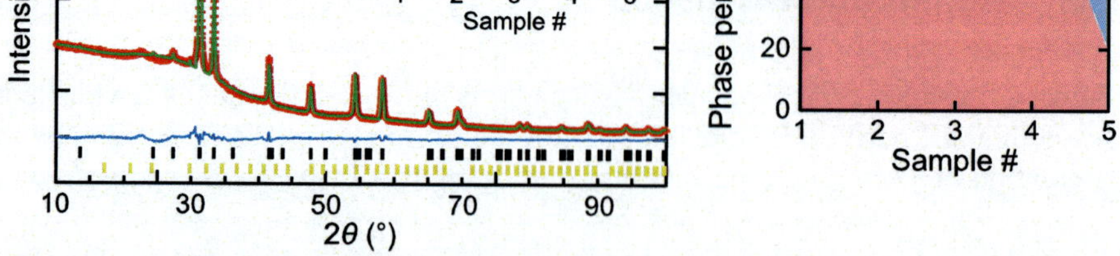

Fig. C.1 **a** Typical powder XRD pattern of Lu_2O_2Bi sample #3 with Rietveld refinement. Inset shows the variation of c/a. **b** Phase diagram for Lu_2O_2Bi samples

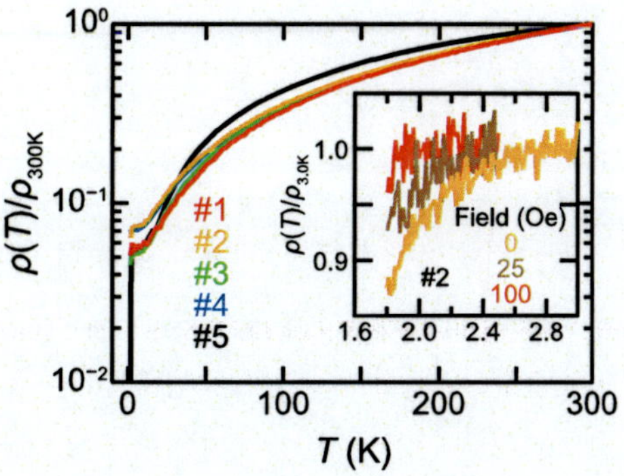

Fig. C.2 Normalized ρ-T curves of Lu_2O_2Bi samples. Inset shows the data at low temperature for sample #2

Lu_2O_2Bi samples were larger than the superconducting threshold value of 3.42 even without O incorporation.

Figure C.2 shows normalized ρ-T curves of Lu_2O_2Bi samples. The Lu_2O_2Bi samples showed a metallic behavior in the entire temperature range similar to those in Y_2O_2Bi and Dy_2O_2Bi. The zero-resistance was observed in O incorporated sample #5 with T_c of 2.16 K. In addition, the ρ drops around 2 K were observed in all other samples and disappeared by applying magnetic fields as shown in the inset of Fig. C.2, indicating the superconducting transition. These results were consistent with the larger c/a values of all Lu_2O_2Bi samples than the threshold value of 3.42.

C.4 Conclusion

$ThCr_2Si_2$-type Lu_2O_2Bi was synthesized for the first time in the form of polycrystalline powder. The O incorporation was achieved by mixing CaO during the synthesis similar to the case of Tb_2O_2Bi, although the c/a value of Lu_2O_2Bi was relatively large even in an ideal composition. This intrinsic enhanced two-dimensionality of Bi^{2-} square net in Lu_2O_2Bi resulted in the emergence of superconductivity in all samples.

References

1. Kim H, Kang CJ, Kim K, Shim J, Min B (2016) Phys Rev B 93(12):125116
2. Izumi F, Momma K (2007) Applied crystallography XX, solid state phenomena, vol 130, pp 15–20. Trans Tech Publications. https://doi.org/10.4028/www.scientific.net/SSP.130.15

Appendix D
Fabrication of Dy_2O_2Bi Thin Film

D.1 Experimental

Dy_2O_2Bi epitaxial thin films were synthesized by the multilayer SPE method (Sect. 3.3). The base and working pressure of the sputtering chamber were less than 3.3×10^{-5} and 2.7 Pa, respectively. The multilayered precursor was composed of Bi (4Nup), Dy (3N), and Dy_2O_3 (3N) layers, deposited by DC mode (20 W) for 11 s, DC mode (50 W) for 16 s, and RF mode (100 W) for 90 s, respectively. The [Dy_2O_3/Dy/Bi/Dy/Bi] unit was deposited on cubic CaF_2 (001) single-crystal substrate (lattice mismatch: 0.34%) with 15 periods. The precursor was *in situ* heated with a heating rate of 30°C / min up to 830°C, and then the temperature was held for 10 min. In this appendix, the film thickness was not evaluated although I assumed it as 50 nm for analysis. Crystal structure was evaluated by XRD using Cu $K\alpha_1$ radiation. Electronic transport properties were measured by the standard four-probe method by PPMS equipped with the sample rotator system. Magnetic properties were measured by a SQUID magnetometer.

D.2 Crystal Structure

Figure D.1a shows XRD out-of-plane 2θ-θ scan of the fabricated film. Dy_2O_2Bi $00l$ diffraction peaks were clearly observed along substrate peaks. In addition, Dy_2O_2Bi 103 diffraction peak was observed in the same ϕ-plane with CaF_2 111 diffraction peak (Fig. D.1b). These results confirmed single crystalline growth of Dy_2O_2Bi thin film on CaF_2 substrate with the epitaxial relationships as Dy_2O_2Bi [100] ll CaF_2 [110] and Dy_2O_2Bi [001] ll CaF_2 [001], similar to that in Y_2O_2Bi epitaxial thin film. Calculated lattice constants were $a = 3.876$ Å and $c = 13.115$ Å. The smaller c-axis length compared with the polycrystalline sample ($c = 13.233$ Å) [1] indicated the incidental F substitution for O site from the substrate, as observed in Y_2O_2Bi

R. Sei, *Two-Dimensional Superconductivity in Rare Earth Oxybismuthides with Unusual Valent Bismuth Square Net*, Springer Theses,
https://doi.org/10.1007/978-981-99-7313-2

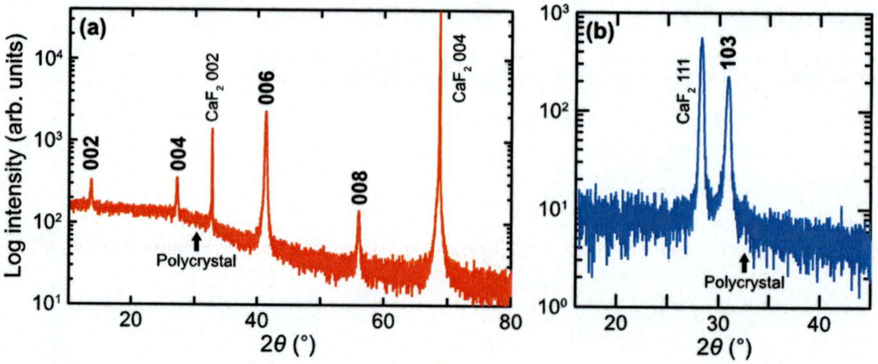

Fig. D.1 **a** XRD out-of-plane 2θ-θ scan of Dy_2O_2Bi epitaxial thin film on the CaF_2 substrate. **b** XRD in-plane 2θ-θ scan around 103 peak

epitaxial thin film. The FWHM of Dy_2O_2Bi 006 rocking curve was 0.44°, indicating the higher crystallinity than Y_2O_2Bi epitaxial thin film.

In the current Dy_2O_2Bi epitaxial thin film, there were small peaks originating from the polycrystalline component of Dy_2O_2Bi phase.[2] These minor peaks disappeared during the sample treatment in the air, indicating the instability of Dy_2O_2Bi phase in the air and surface oxidation of the sample. These surface residues might have had some influence on the physical properties of the film.

D.3 Physical Properties

Figure D.2a shows ρ-T curve of Dy_2O_2Bi epitaxial thin film. The film showed similar metallic behavior to Y_2O_2Bi epitaxial thin film. The MR showed a small positive value at 9 T in all temperature ranges. Figure D.2b shows χ-T curves where magnetic fields were applied parallel and perpendicular to c-axis. Both curves were fitted by the Curie–Weiss law, but a small deviation was observed at ~7 K (inset of Fig. D.2b). This was attributed to the AF transition of $[Dy_2O_2]^{2+}$ slab, which was consistent with T_{N1} of F doped polycrystalline Dy_2O_2Bi (Appendix B). A continuous rise of χ below T_{N1} was probably due to the surface residue. This may have hidden the small AF signal around 40 K observed in F doped polycrystalline Dy_2O_2Bi (T_{N2} in Appendix B).

Figure D.2c shows MR at 2 K. The sudden MR drop around the zero field was attributed to the weak localization effect originating from the disorder of the film such as polycrystalline component and surface residue. The degree of MR value

[2] Pure Dy_2O_2Bi epitaxial thin film was obtained by changing the fabrication condition: multilayer period from 15 to 14 times, and heating time from 10 min to 9 min 20 s. However, this sample was oxidized during the soldering for the electrical measurement.

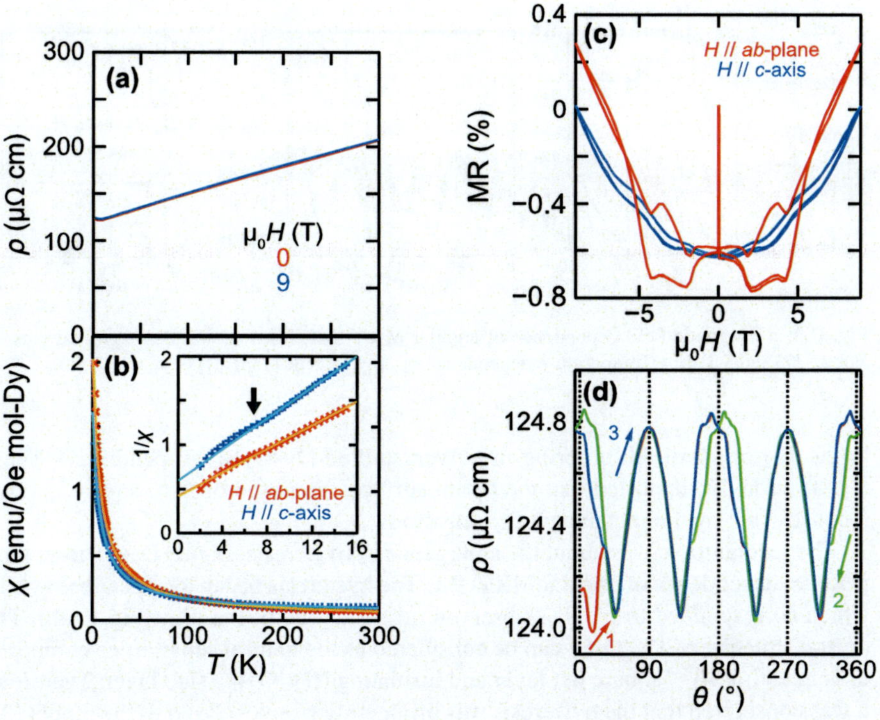

Fig. D.2 Physical properties of Dy_2O_2Bi epitaxial thin film. **a** ρ-T curve at 0 and 9 T. **b** χ-T curves with different magnetic field directions. Solid lines denote Curie–Weiss fitting and the black arrow denotes the T_{N1}. **c** MR data with different magnetic field directions. **d** Angular dependence of MR at 2 K and 9 T. Measurement configuration was the same as that of Y_2O_2Bi epitaxial thin film

was similar to the result of F doped polycrystalline Dy_2O_2Bi (Appendix B). In addition, a four-fold symmetric component appeared in angular dependence of MR (Fig. D.2d), which was a contrast to the two-fold behavior in Y_2O_2Bi demonstrating two-dimensional electrical transport. Notably, the hysteresis curve was observed in both MR at 2 K and angular dependence of MR at 2 K and 9 T. Hereafter, I discussed the reason of these novel phenomena.

First of all, similar angular dependence of MR is observed in $Ca_{2-x}Sr_xRuO_4$, where the two-fold symmetric behavior turns to four-fold symmetric one by increasing applied magnetic field [2]. This is attributed to the Fermi surface reconstruction driven by metamagnetic transition. In order to verify this situation, field and temperature dependence of angular MR were evaluated. Figure D.3a shows field dependence of angular MR at 2 K. The MR at 1 T showed two-fold symmetric behavior; on the other hand, the four-fold symmetric component began to grow at 3 T and more. This field approximately corresponded to the critical field of the metamagnetic transition of polycrystalline Dy_2O_2Bi (Appendix B). In addition, Fig. D.3c indicates that the four-fold symmetric component was observed below 50 K, which corresponded

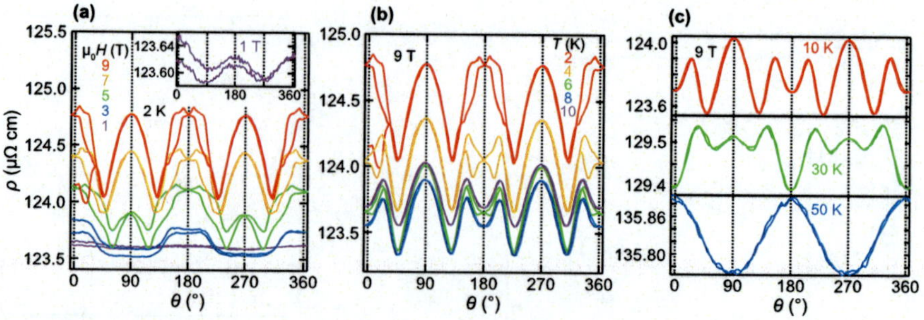

Fig. D.3 **a** Magnetic field dependence of angular MR at 2 K. Inset shows the magnified view of MR at 2 K and 9 T. **b, c** Temperature dependence of angular MR at 9 T

to the T_{N2} of small AF ordering in polycrystalline Dy_2O_2Bi (Appendix B). These results strongly supported that the Fermi surface composed of Bi^{2-} square net was reconstructed by the metamagnetic transition.

The second topic was about the emergence of hysteresis. Figure D.3b shows temperature dependence of angular MR at 9 T. The hysteresis began to appear below 8 K, which corresponded to the T_{N1} observed in χ-T curve of the film (Fig. D.2b). The crystal structure of Dy_2O_2Bi can be considered as the natural superlattice composed of conducting Bi^{2-} square net layer and insulating $[Dy_2O_2]^{2+}$ slab layer. Therefore, it was considered that the hysteresis was originating from one kind of the tunnel MR effect when carriers conducted between inter Bi^{2-} square nets through the AF or metamagnetic ordered $[Dy_2O_2]^{2+}$ slab.

D.4 Conclusion

Dy_2O_2Bi epitaxial thin film was successfully obtained by the multilayer SPE method. This result demonstrated the usefulness of this novel fabrication technique to investigate the physical properties of R_2O_2Bi compounds. Dy_2O_2Bi epitaxial thin film showed unusual magnetoresistance probably originating from the complicated interaction between Bi^{2-} square net and magnetic $[Dy_2O_2]^{2+}$ slab. This result clearly indicates that R substitution was a powerful means to induce various physical properties in R_2O_2Bi compounds.

References

1. Nuss J, Jansen M (2012) Zeitschrift für anorganische und allgemeine Chemie 638(3–4):611
2. Balicas L, Nakatsuji S, Hall D, Ohnishi T, Fisk Z, Maeno Y, Singh DJ (2005) Phys Rev Lett 95:196407